Great Science Adventures

Discovering Earth's Landforms and Surface Features

Dinah Zike
and
Susan Simpson

Great Science Adventures is a comprehensive project which is projected to include the titles below. Please check our website, www.greatscienceadventures.com, for updates and product availability.

Great Life Science Studies:
The World of Plants
The World of Insects and Arachnids
Discovering the Human Body and Senses
The World of Vertebrates
Discovering Biomes - Earth's Ecosystems
The World of Health and Safety

Great Physical Science Studies:
The World of Tools and Technology
The World of the Light and Sound
Discovering Atoms, Molecules, and Matter
Discovering Energy, Forces, and Motion
Discovering Magnets and Electricity

Great Earth Science Studies:
The World of Space
Discovering Earth's Landforms and Surface Features
Discovering Earth's Atmosphere and Weather
Discovering Rocks and Minerals
Discovering Earth's Oceans and Fresh Water

Copyright © 2002 by:
Common Sense Press
8786 Highway 21
P.O. Box 1365
Melrose, FL 32666
(352) 475–5757
www.greatscienceadventures.com

Printed in the United States of America
ISBN 1-929683-12-X

The authors and the publisher have made every reasonable effort to ensure that the experiments and activities in this book are safe when performed according to the book's instructions. We assume no responsibility for any damage sustained or caused while performing the activities or experiments in *Great Science Adventures*. We further recommend that students undertake these activities and experiments under the supervision of a teacher, parent, and / or guardian.

Great Science Adventures

Table of Contents

Great Science Adventures

Introduction

Great Science Adventures is a unique, highly effective program that is easy to use for teachers as well as students. This book contains 24 lessons. The concepts to be taught are clearly listed at the top of each lesson. Activities, questions, clear directions, and pictures are included to facilitate learning. Each lesson takes one to three days to complete.

This program utilizes highly effective methods of learning. Students not only gain knowledge of basic science concepts, but also learn how to apply them.

Specially designed *3D Graphic Organizers* are included for use with the lessons. These organizers review the science concepts while adding to your students' understanding and retention of the subject matter.

This *Great Science Adventures* book is divided into four parts:

1) Following this *Introduction* you will find the *How to Use This Program* section. It contains all the information you need to make the program successful. The *How to Use This Program* section also contains instructions for Dinah Zike's *3D Graphic Organizers*. Please take the time to learn the terms and instructions for these learning manipulatives.

2) In the *Teacher's Section,* the numbered lessons include a list of the science concepts to be taught, simple to complex vocabulary words, and activities that reinforce the science concepts. Each activity includes a list of materials needed, directions, pictures, questions, written assignments, and other helpful information for the teacher.

   The *Teacher's Section* also includes enrichment activities entitled *Experiences, Investigations, and Research.* Alternative assessment suggestions are found at the end of the *Teacher's Section.*

3) The *Lots of Science Library Books* are next. These books are numbered to correlate with the lessons. Each *Lots of Science Library Book* will cover all the concepts included in its corresponding lesson. You may read the *LSLB* to your students, ask them to read the books on their own, or make the books available as research materials. Covers for the books are found at the beginning of the *LSLB* section. (Common Sense Press grants permission for you to photocopy the *Lots of Science Library Books* pages and covers for your students.)

4) *Graphics Pages,* also listed by lesson numbers, provide pictures and graphics that can be used with the activities. They can be duplicated and used on student–made manipulatives or students may draw their own illustrations. The *Investigative Loop* at the front of this section may be photocopied as well. (Common Sense Press grants permission for you to photocopy the *Graphics Pages* for your students.)

How to Use This Program

This program can be used in a single level classroom, multilevel classroom, homeschool, co–op group, or science club. Everything you need for a complete Earth study is included in this book. Intermediate students will need access to basic reference materials.

Take the time to read the entire *How to Use this Program* section and become familiar with the sections of this book described in the *Introduction*.

Begin a lesson by reading the *Teacher Pages* for that lesson. Choose vocabulary words for each student and the activities to complete. Collect the materials you need for these activities.

Introduce the lesson with the *Lots of Science Library Book* by reading it aloud or by asking a student to read it. (The *Lots of Science Library Books* are located after the *Teacher's Section* in this book.)

Discuss the concepts presented in the *Lots of Science Library Book,* focusing on the ones listed in your *Teacher's Section*.

Follow the directions for the activities you have chosen.

How to Use the Multilevel Approach

The lessons in this book include basic content appropriate for grades K–8 at different mastery levels. For example, throughout the teaching process, a first grader will be exposed to a lot of information but would not be expected to retain all of it. In the same lesson, a sixth grade student will learn all the steps of the process, be able to communicate them in writing, and be able to apply that information to different situations.

In the *Lots of Science Library Books,* the words written in larger type are for all students. The words in smaller type are for upper level students and include more scientific details about the basic content, as well as interesting facts for older learners.

In the activity sections, icons are used to designate the levels of specific writing assignments.

This icon ✎ indicates the Beginning level, which includes the non reading or early reading student. This level applies mainly to kindergarten and first grade students.

This icon ✎✎ is used for the Primary level. It includes the reading student who is still working to be a fluent reader. This level is designed primarily for second and third graders.

This icon ✎✎✎ denotes the Intermediate level, or fluent reader. This level of activities will usually apply to fourth through eighth grade students.

If you are working with a student in seventh or eighth grade, we recommend using the assignments for the Intermediate level, plus at least one *Experiences, Investigations, and Research* activity per lesson.

No matter what grade level your students are working on, use a level of written work that is appropriate for their reading and writing abilities. It is good for students to review data they already know, learn new data and concepts, and be exposed to advanced information and processes.

Vocabulary Words

Each lesson contains a list of vocabulary words used in the content of the lesson. Some of these words will be "too easy" for your students, some will be "too hard," and others will be "just right." The "too easy" words will be used automatically during independent writing assignments. Words that are "too hard" can be used during discussion times. Words that are "just right" can be studied by definition, usage, and spelling. Encourage your students to use these words in their own writing and speaking.

You can encourage beginning students to use their vocabulary words as you reinforce reading instruction and enhance discussions about the topic, and as words to be copied in cooperative writing, or teacher guided writing.

Primary and Intermediate students can make a Vocabulary Book for new words. Instructions for making a Vocabulary Book are found on page 3. The Vocabulary Book will contain the word definitions and sentences composed by the student for each word. Students should also be expected to use their vocabulary words in discussions and independent writing assignments. A vocabulary word with an asterisk (*) next to it is designated for Intermediate students only.

Using 3D Graphic Organizers

The *3D Graphic Organizers* provide a format for students of all levels to conceptualize, analyze, review, and apply the concepts of the lesson. The *3D Graphic Organizers* take complicated information and break it down into visual parts so students can better understand the concepts. Most *3D Graphic Organizers* involve writing about the subject matter. Although the content for the levels will generally be the same, assignments and expectations for the levels will vary.

Beginning students may dictate or copy one or two "clue" words about the topic. These students will use the written clues to verbally communicate the science concept. The teacher should provide various ways for the students to restate the concept. This will reinforce the science concept, and encourage the students in their reading and higher order thinking skills.

Primary students may write or copy one or two "clue" words and write a sentence about the topic. The teacher should encourage students to use vocabulary words when writing these sentences. As students read their sentences and discuss the concept, they will reinforce the science concept, increasing their fluency in reading, and higher order thinking skills.

Intermediate students may write several sentences or a paragraph about the topic. These students are also encouraged to use reference materials to expand their knowledge of the subject. As tasks are completed, students enhance their abilities to locate information, read for content, compose sentences and paragraphs, and increase vocabulary. Encourage these students to use the vocabulary words in a context that indicates understanding of the words' meanings.

Illustrations for the *3D Graphic Organizers* are found on the *Graphics Pages* and are labeled by the lesson number and a letter, such as 5–A. Your students may either use these graphics to draw their own pictures, or cut out and glue them directly on their work.

Several of the *3D Graphic Organizers* expand over a series of lessons. For this reason, you will need a storage system for each students' *3D Graphic Organizers*. A pocket folder or a reclosable plastic bag works well. See page 1 for more information on storing materials.

Investigative Loop™

The *Investigative Loop* is used throughout *Great Science Adventures* to ensure that your labs are effective and practical. Labs give students a context for the application of their science lessons so they can begin to take ownership of the concepts, increasing understanding as well as retention.

The *Investigative Loop* can be used in any lab. The steps are easy to follow, user friendly, and flexible.

Each *Investigative Loop* begins with a **Question or Concept.** If the lab is designed to answer a question, use a question in this phase. For example, the question could be: "How much air is in soil?" Since the activity for this lab will show the amount of air in the soil, a question is the best way to begin this *Investigative Loop*.

If the lab is designed to demonstrate a concept, use a concept statement in this phase, such as: "The freeze-thaw process affects clay." The lab will demonstrate that fact to the students.

After the **Question or Concept** is formulated, the next phase of the *Investigative Loop* is Research and/or Predictions. Research gives students a foundation for the lab. Having researched the question or concept, students enter the lab with a basis for understanding what they observe. Predictions are best used when the first phase is a question. Predictions can be in the form of a statement, a diagram, or a sequence of events.

The **Procedure** for the lab follows. This is an explanation of how to set up the lab and any tasks involved in it. A list of materials for the lab may be included in this section or may precede the entire *Investigative Loop*.

Whether the lab is designed to answer a question or demonstrate a concept, the students' **Observations** are of prime importance. Instruct the students concerning what they are to focus upon in their observations. The Observation phase will continue until the lab ends.

Once observations are made, students must **Record the Data**. Data may be recorded through diagrams or illustrations. Recording quantitative or qualitative observations about the lab is another important activity in this phase. Records may be kept daily for an extended lab or at the beginning and end for a short lab.

Conclusions and/or Applications are completed when the lab ends. Usually the data records will be reviewed before a conclusion can be drawn about the lab. Encourage the students to defend their conclusions by using the data records. Applications are made by using the conclusions to generalize to other situations or by stating how to use the information in daily life.

Next we must **Communicate the Conclusions**. This phase is an opportunity for students to be creative. Conclusions can be communicated through a graph, story, report, video, mock radio show, etc. Students may also participate in a group presentation.

Questions that are asked as the activity proceeds are called **Spark Questions.** Questions that the lab sparks in the minds of the students are important to follow and discuss when the lab ends. The lab itself will answer many of these questions, while others may lead to a new *Investigative Loop*. Assign someone to keep a list of all Spark Questions.

One lab naturally leads to another. This begins a new *Investigative Loop*. The phase called **New Loop** is a brainstorming time for narrowing the lab down to a new question or concept. When the new lab has been decided upon, the *Investigative Loop* begins again with a new Question or Concept.

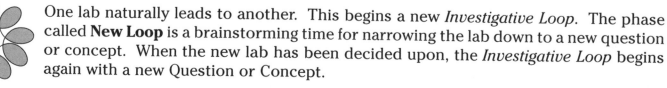

Take the time to teach your students to make qualitative and quantitative observations. Qualitative observations involve recording the color, texture, shape, smell, size (as small, medium, large), or any words that describe the qualities of an object. Quantitative observations involve using a standard unit of measurement to determine the length, width, weight, mass or volume of an object.

All students will make a Lab Book, in the form of a Large Question and Answer Book, to record information about the Investigative Loops. Instructions are found on page 2. Your students will make a new Lab Book as needed to glue side–by–side to the previous one. Instructions can be found in the *Teacher's Section*.

Predictions, data, and conclusions about the *Investigative Loops* are written under the tabs in this book.

When you begin an *Investigative Loop*, ask your students to glue or draw the graphic of the experiment on the tab of the Lab Book. Each *Investigative Loop* is labeled with the lesson number and another number. These numbers are also found on the corresponding graphics.

During an *Investigative Loop*, beginning students should be encouraged to discuss their answers to all experiment questions. By discussing the topic, the students will not only learn the science concepts and procedures, but will be able to organize their thinking in a manner that will assist them in later years of writing. This discussion time is very important for beginning students and should not be rushed.

After the discussion, work with the students to construct a sentence about the topic. Let them copy the sentence. Students can also write "clue" words to help them remember key points about the experiment for discussion at a later time.

Primary students should be encouraged to verbalize their answers. By discussing the topic, students will learn the science concepts and procedures and learn to organize their thinking, increasing their ability to use higher level thinking skills. After the discussion, students can complete the assignment using simple phrases or sentences. Encourage students to share the information they have learned with others, such as parents or friends. This will reinforce the content and skills covered in the lesson.

Even though Intermediate students can write the answers to the lab assignments, the discussion process is very important and should not be skipped. By discussing the experiments, students review the science concepts and procedures as well as organize their thinking for the writing assignments. This allows them to think and write at higher levels. These students should be encouraged to use their vocabulary words in their lab writing assignments.

Design Your Own Experiment

After an *Investigative Loop* is completed, intermediate students have the option to design their own experiments based on that lab. The following procedure should be used for those experiments.

Select a Topic based upon an experience in an *Investigative Loop*, science content, an observation, a high-interest topic, a controversial topic, or a current event.

Discuss the Topic as a class, in student groups, and with knowledgeable professionals.

Read and Research the Topic using the library, the Internet, and hands-on investigations and observations, when possible.

Select a Question that can be investigated and answered using easily obtained reference materials, specimens, and/or chemicals, and make sure that the question selected lends itself to scientific inquiry. Ask specific, focused questions instead of broad, unanswerable questions. Questions might ask "how" something responds, influences, behaves, determines, forms, or is similar or different to something else.

Predict the answer to your question, and be prepared to accept the fact that your prediction might be incorrect or only partially correct. Examine and record all evidence gathered during testing that both confirms and contradicts your prediction.

Design a Testing Procedure that gathers information that can be used to answer your question. Make sure your procedure results in empirical, or measurable, evidence. Don't forget to do the following:

> Determine where and how the tests will take place – in a natural (field work) or controlled (lab) setting.

> Collect and use tools to gather information and enhance observations.
> Make accurate measurements. Use calculators and computers when appropriate.

> Plan how to document the test procedure and how to communicate and display resulting data.

> Identify variables, or things that might prevent the experiment from being "fair." Before beginning, determine which variables have no effect, a slight effect, or a major effect on your experiment. Create a method for controlling these variables.

Conduct the Experiment carefully and record your findings.

Analyze the Question Again. Determine if the evidence obtained and the scientific explanations of the evidence are reasonable based upon what is known, what you have learned, and what scientists and specialists have reported.

Communicate Findings so that others can duplicate the experiment. Include all pertinent research, measurements, observations, controls, variables, graphs, tables, charts, and diagrams. Discuss observations and results with relevant people.

Reanalyze the Problem and if needed, redefine the problem and retest. Or, try to apply what was learned to similar problems and situations.

Experiences, Investigations, and Research

At the end of each lesson in the *Teacher's Section* is a category of activities entitled *Experiences, Investigations, and Research*. These activities expand upon concepts taught in the lesson, provide a foundation for further study of the content, or integrate the study with other disciplines. The following icons are used to identify the type of each activity.

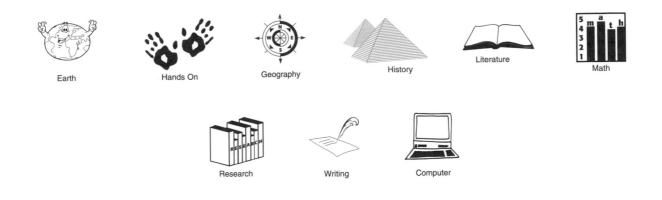

Earth   Hands On   Geography   History   Literature   Math

Research   Writing   Computer

Cumulative Project

At the end of the program we recommend that students compile a Cumulative Project using the activities they have completed during their course of study. It may include the Investigative Loops, Lab Book, and the 3D Graphic Organizers on display.

Please do not overlook the Cumulative Project, as it provides immeasurable benefits for your students. Students will review all the content as they create the project. Each student will organize the material in his or her unique way, providing an opportunity for authentic assessment and for reinforcing the context in which it was learned. This project creates a format where students can make sense of the whole study in a way that cannot be accomplished otherwise.

Fast Food and Fast Folds

"If making the manipulatives takes up too much of your instructional time, they are not worth doing. They have to be made quickly, and they can be, if the students know exactly what is expected of them. Hamburgers, Hot Dogs, Tacos, Mountains, Valleys, and Shutter–Folds can be produced by students, who in turn use these folds to make organizers and manipulatives."– Dinah Zike

Every fold has two parts. The outside edge formed by a fold is called the **"Mountain."** The inside of this edge is the **"Valley."**

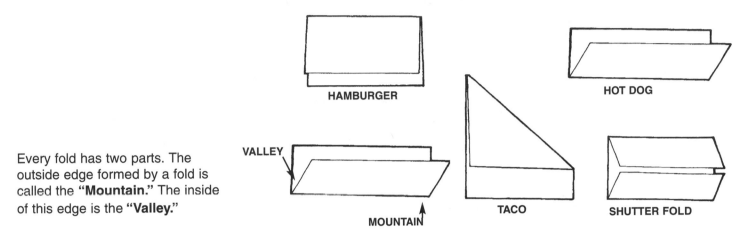

HAMBURGER

HOT DOG

VALLEY

MOUNTAIN

TACO

SHUTTER FOLD

Storage – Book Bags

One–gallon reclosable plastic bags are ideal for storing ongoing projects and books that students are writing and researching.

Use strips of clear, 2" tape to secure 1" x 1" pieces of index card to the front and back of one of the top corners of a bag, under the closure. Punch a hole through the index cards. Use a giant notebook ring to keep several of the "Book Bags" together.

Label the bags by writing on them with a permanent marker.

Alternatively, the bags can be stored in a notebook if you place the 2" clear tape along the side of the storage bag and punch 3 holes in the tape.

Half Book

Fold a sheet of paper in half like a Hamburger.

HAMBURGER

Pocket Book

1. Fold a sheet of paper in half like a Hamburger.

2. Open the folded paper and fold one of the long sides up two and a half–inch inches to form a pocket. Refold along the Hamburger fold so that the newly formed pockets are on the inside.

3. Glue the outer edges of the two and a half–inch fold with a small amount of glue.

4. Make a multi–paged booklet by gluing several Pocket Books "side–by–side."

5. Glue a construction paper cover around the multi–page pocket booklet.

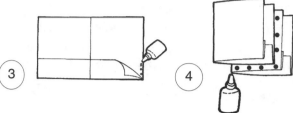

Side–by–Side

Some books can easily grow into larger books by gluing them side–by–side. Make two or more of these books. Be sure the books are closed, then glue the back cover of one book to the front cover of the next book. Continue in this manner, making the book as large as needed. Glue a cover over the whole book.

Vocabulary Book

1. Take two sheets of paper and fold each sheet like a Hot Dog.

2. Fold each Hot Dog in half like a Hamburger. Fold each Hamburger in half two more times and crease well. Unfold the sheets of paper, which are divided into sixteenths.

3. On one side only, cut the folds up to the Mountain top, forming eight tabs. Repeat this process on the second sheet of paper.

4. Take a sheet of construction paper and fold like a Hot Dog. Glue the back of one vocabulary sheet to one of the inside sections of the construction paper. Glue the second vocabulary sheet to the other side of the construction paper fold.

5. Vocabulary Books can be made larger by gluing them "side–by–side."

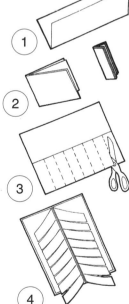

The *Lots of Science Library Book* Shelf

Make a bookshelf for the *Lots of Science Library Books* using an appropriate sized box or following the instructions below.

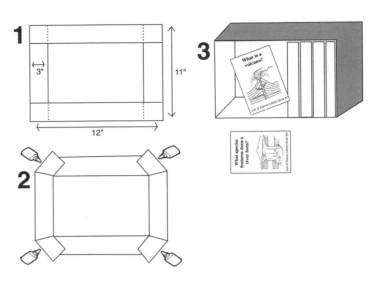

1. Begin with an 11" x 12" piece of poster board or cardboard. Mark lines 3" from the edge of each side. Fold up on each folded line. Cut on the dotted lines as indicated in illustration 1. Refold on the line.

2. Glue the tabs under the top and bottom sections of the shelf. See illustration 2.

3. If you are photocopying your *Lots of Science Library Books*, consider using green paper for the covers and the same green paper to cover your bookshelf.

Large Question and Answer Book

1. Fold a sheet of paper in half like a Hamburger. Fold it in half again like a Hamburger. Cut up the valley of the inside fold, forming two tabs.

2. A larger book can be made by gluing Large Question and Answer Books "side by side."

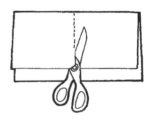

Small Question and Answer Book

1. Fold a sheet of paper in half like a Hot Dog.

2. Fold this long rectangle in half like a Hamburger.

3. Fold both ends back to touch the Mountain Top.

4. On the side forming two Valleys and one Mountain Top, make vertical cuts through one thickness of paper, forming tabs for questions and answers. These four tabs can also be cut in half making eight tabs.

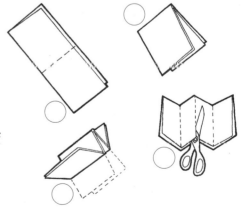

Layered Look Book

1. Stack two sheets of paper and place the back sheet one inch higher than the front sheet.

2. Bring the bottom of both sheets upward and align the edges so that all of the layers or tabs are the same distance apart.

3. When all tabs are an equal distance apart, fold the papers and crease well.

4. Open the papers and glue them together along the Valley/center fold.

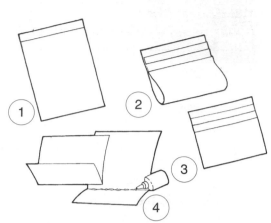
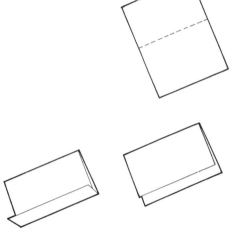

Matchbook

1. Fold a sheet of paper like a hamburger, but fold it so that one side is one inch longer than the other side.

2. Fold the one inch tab over the short side forming an envelope-like fold.

Teacher's Section

STUDY A GREAT EARTH SCIENCE

Lithosphere/Earth Concept Map

Lessons 1-8

Numbers refer to Lesson Numbers

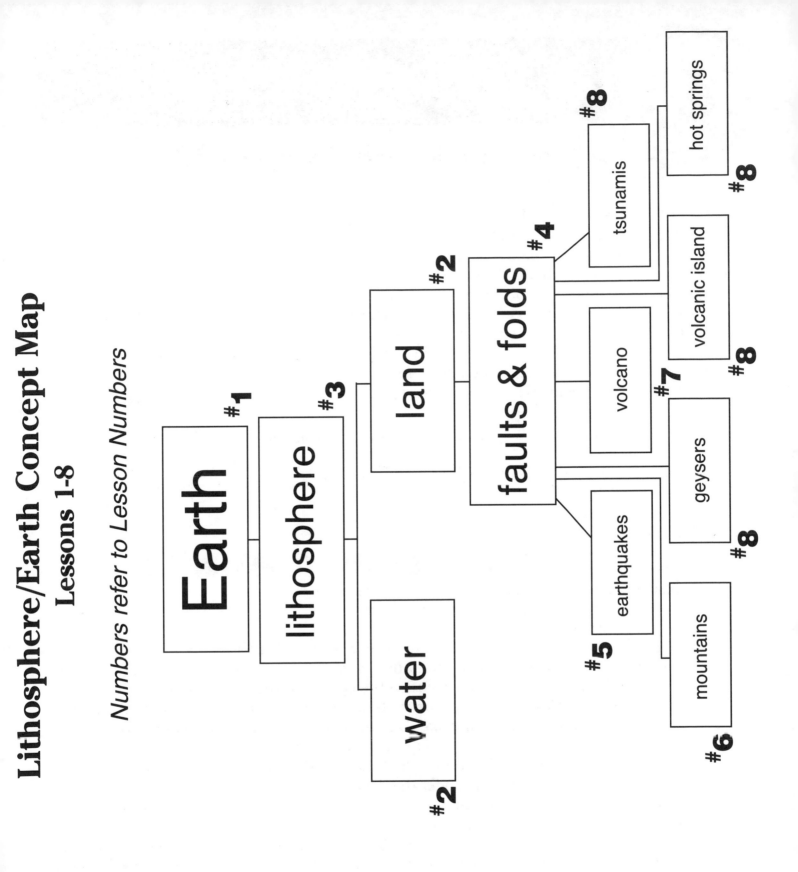

Where is Earth in space?

Lithosphere Concepts:

- Earth is located in the Milky Way Galaxy.
- It is the third planet from the Sun.
- Its revolution takes about 365.25 days, and its rotation takes about 24 hours.
- Earth is 7,926 miles (12,753 km) in diameter.
- Earth's structure consists of a solid metal inner core surrounded by a liquid metal outer core, a rocky mantle, and a thin, rocky crust.
- Earth is composed of many different elements, but some of the most common are iron, nickel, silicon, magnesium, aluminum, and oxygen.
- Earth is the only planet known to support life.

Vocabulary: Earth   Sun   planet   *revolution   *rotation

Read: *Lots of Science Library Book #1.*

Activities:

Planet Earth – Graphic Organizer

Focus Skills: communicating information   recording data

Paper Handouts: 12" x 18" sheet of construction paper   2 sheets of 8.5" x 11" paper
a copy of Graphics 1A-E

Graphic Organizer: Use the 12" x 18" paper to make a Shutter Fold. Cut Graphic 1A on the dotted line and glue it to the top front of the Shutter Fold Project. Title it *Earth.* Fold the 8.5" x 11" paper in a Taco and cut off the end. Fold the Taco into a Hamburger and cut on the fold. Make two Matchbooks. Glue Graphic 1B on one and 1C on the other Matchbook. Label them *Earth in the Solar System* and *Earth rotates on its axis,* accordingly. On the top section inside each Matchbook:

✎ Draw the solar system and circle Earth. Write *365 days* under the drawing. Draw Earth on its axis, and write *24 hours* under the drawing.

✎✎ Write clue words about each Graphic: *third planet, solid, supports life, 365 days to orbit and rotates on axis, 24 hours, day and night.*

✎✎✎ Explain Earth's location in space. Describe its revolution and size. Explain the rotation of Earth and why it gives us day and night. Include the *Fascinating Physical Features of Earth* from the *Lots of Science Library Book #1* in your explanation. Illustrate as needed.

Glue these two Matchbooks to the middle of each front tab on the *Earth Shutter Fold Project.*

Label and color Graphics 1D and E. Glue them inside the Matchbooks from the previous activity on the *Earth Shutter Fold Project.*

Day and Night

Activity Materials: ball or globe   flashlight

Activity: Hold the ball or globe at the top and bottom. Slowly rotate it in the same manner that Earth rotates on its axis. Ask a partner to hold a flashlight directly in line with the ball or globe, shining light on it. Explain how this activity illustrates day and night on Earth.

Earth and the Sun

Paper Handouts: scrap paper   newspaper   optional: crayons and yellow paint

Activity: Cut a circle 1/4" in diameter.. Color it green and blue. Use newspaper to make a 22" in diameter circle. Paint it yellow. Place the two circles 4 3/4" (12.1 cm) apart. This is a scale model of the Sun and Earth.

Note: To complete the entire solar system, refer to the chart below.

Planet	Diameter of Circle	Distance from paper Sun
Mercury	1/8"	1 3/4" (4.4 cm)
Venus	1/4"	3 1/4" (8.3 cm)
Earth	1/8"	4 3/4" (12.1 cm)
Mars	1/4"	7" (17.8 cm)
Jupiter	2 3/4"	2' (61 cm)
Saturn	2 3/8"	3' 8" (111.8 cm)
Uranus	1"	7' 5" (226.1 cm)
Neptune	7/8"	11' 8" (355.6 cm)
Pluto	1/8"	15' 3" (464.8 cm)

Experiences, Investigations, and Research

Select one or more of the following activities for individual or group enrichment projects. Allow your students to determine the format in which they would like to report, share, or graphically present what they have discovered. This should be a creative investigation that utilizes your students' strengths.

1. Use a compass to observe the magnetic field of Earth. The needle of a compass will always point towards Earth's magnetic north pole.

2. Read *Journey to the Center of the Earth* by Jules Verne.

3. Compare and contrast a boiled egg to the Earth inside and out. How do the shapes differ? Compare Earth's crust to the egg's shell. Could the egg's yolk represent the Earth's inner and outer metal cores, and the egg white the mantle?

4. Estimate and graph what percentage you think Earth's crust, mantle and core would represent by volume. For example, Earth's crust is less than 1% of the total volume of Earth. Research the thickness of the mantle and the core, and use this information to estimate their volume percentages.

5. Use clay to model the concentric layers of Earth: inner core, outer core, mantle, and crust.

6. NASA Earth Oberservatory - http://earthobservatory.nasa.gov (Click on "The Blue Marble".)

7. The Milky Way Galaxy - http://casswww.ucsd.edu/public/tutorial/MW.html

What are the physical features of Earth?

Lithosphere Concepts:

- Earth's surface is about 79% water and 21% land.
- Earth has a protective atmosphere surrounding it.
- Imaginary lines are used to locate positions on Earth's surface.
- Physical features of Earth's surface include mountains, valleys, volcanoes, plains, plateaus, canyons, and caves.

Vocabulary: ocean   continent   *atmosphere

Read: *Lots of Science Library Book #2.*

Activities:

Land and Water – Graphic Organizer

Focus Skill: graphing
Paper Handouts: a copy of Graphics 2A-B   *Earth Shutter Fold Project*
Graphic Organizer: Color and cut out Graphics 2A-B. Cut up on the middle dotted line in each circle. Insert the cut-up line from one circle into the other one, so the circles can be turned. Move the circles so that they show a graph indicating the percentage of water (blue) and land (green) on Earth's surface. When the circles are aligned properly, glue them together in place. Glue this graph on the front or back of the *Earth Shutter Fold Project*. Write, copy, or dictate an explanation of the graph, such as *Earth's surface is about 79% water and 21% land.*

Earth Map – Graphic Organizer

Focus Skill: recording data
Paper Handouts: a copy of Graphics 2C-D   *Earth Shutter Fold Project*
Graphic Organizer: Glue Graphics 2C-D on the inside of the Earth Shutter Fold Project. Draw and label these lines: *equator, Tropic of Cancer, Tropic of Capricorn, and Prime Meridian.* Refer to *Lots of Science Library Book #2.*

Fascinating Facts about Earth – Graphic Organizer

Focus Skill: map reading
Paper Handouts: a copy of Graphic 2E   *Earth Shutter Fold Project*
Graphic Organizer: Cut out Graphic 2E and fold it on the middle line so that the illustration of water is on the cover. This represents the Pacific Ocean. Follow the directions below for the inside and glue it to the appropriate place on the world map inside the *Earth Shutter Fold Project.*

✏️ Draw the cover picture on the inside and color it.

✏️✏️ Copy information from the *Lots of Science Library Book* about the Pacific Ocean: *world's largest ocean.*

✏️✏️✏️ Write information about the cover picture.

Teacher's Note: The Graphics for all the Fascinating Facts activities are grouped together in the Graphic Pages.

Experiences, Investigations, and Research

Select one or more of the following activities for individual or group enrichment projects. Allow your students to determine the format in which they would like to report, share, or graphically present what they have discovered. This should be a creative investigation that utilizes your students' strengths.

 1. Compare and contrast how maps and globes represent the surface of the Earth. Which is more realistic? Why?

2. Locate latitude and longitude lines on a map or a globe. Determine the longitude and latitude for your area.

3. Use a permanent marker to draw an equator on a beach ball, and label its "northern and southern hemispheres." Draw shapes on the ball to represent the seven continents. Label the continents and oceans. Toss the ball up in the air and catch it. Where are your hands? in which hemisphere? on which continent or ocean?

4. Make a graph to show what percentages of Earth's surface are covered in water and land (79% water and 21% land).

5. Write a story about Goldilocks traveling through the solar system looking for a planet that is "just right."

6. How much water is there on and in the Earth?
   http//ga.water.usgs.gov/edu/earthhowmuch.html

7. Earth's Atmosphere
   http://liftoff.msfc.nasa.gov/academy/space/atmosphere.html

Great Science Adventures

What is the lithosphere?

Lithosphere Concepts:

- Lithosphere is Greek for "sphere of stone." At one time the term lithosphere was used to refer to the mantle and crust.
- Today, lithosphere refers to the solid, cool, outermost layer of Earth; the crust and uppermost mantle.
- Under the continents, Earth's crust is about 25 miles thick, and under the ocean floor it is about 5 miles thick.
- Earth's crust is divided into gigantic sections called continental plates.
- These continental plates can move apart, collide, or slide sideways.
- Most scientists think that at one time all the continents were joined together, forming one massive land formation called Pangaea.

Teacher's Note: An alternative assessment suggestion for this lesson is found on pages 64-65. If Graphics Pages are being consumed, first photocopy the assessment graphics that are needed.

Vocabulary: lithosphere   continental plates   *plate tectonics

Read: *Lots of Science Library Book #3.*

Activities:

The Continents – Graphic Organizer

Focus Skills: map reading   applying new data
Paper Handouts: a copy of Graphic 3A
          optional: poster board or thin cardboard
Graphic Organizer: Cut out the continents in Graphic 3A. If desired, glue them on poster board or thin cardboard and cut them out again. Spend time experimenting with the shapes of the continents. Make the land formation Pangaea. Place the continents in their present locations.

Fascinating Facts about Earth – Graphic Organizer

Focus Skill: map reading
Paper Handouts: a copy of Graphic 3B   *Earth Shutter Fold Project*
  Graphic Organizer: Cut out Graphic 3B and fold on the middle line so that the illustration is on the cover. Follow the directions below and glue it to the appropriate place on the world map inside the *Earth Shutter Fold Project.*

✎ Draw the cover picture on the inside and color it.

✎✎ Copy information from the *Lots of Science Library Book* about the cover picture: *hot liquid rock forms new rock.*

✎✎✎ Write information about the cover picture.

Experiences, Investigations, and Research

Select one or more of the following activities for individual or group enrichment projects. Allow your students to determine the format in which they would like to report, share, or graphically present what they have discovered. This should be a creative investigation that utilizes your students' strengths.

1. Investigate the history of the theory of plate tectonics. Include information on continental drift, sea floor spreading, subduction zones, and modern views on geodynamics.

2. Who, What, When, Where: Alfred Wegener, (1880-1930), Germany.

3. Using two stacks of towels, design a demonstration that illustrates the ways that plates move.

4. Research and draw the following: Pangaea dividing into Gondwannaland and Laurasia.

5. Compare and contrast the trenches and ridges found on the ocean floor.

6. Physical Features of the Earth
   http://www.uen.org/utahlink/lp_res/TRB023.html

7. The Grand Canyon and Continental Drift
   http://www.kaibab.org/geology/contdrft.htm

What are faults and folds in the lithosphere?

Lithosphere Concepts:

- When pressure and stresses from the inner energy of the Earth and the movement of continental plates build up, rock layers can break, creating faults.
- Faults are cracks in the lithosphere. Major faults are located at continental plate boundaries.
- Sometimes rocks do not break when pressure builds up underneath them. Instead, these rock layers bend or crumple resulting in folds.

Vocabulary: fault   fold   *anticline   *syncline

Read: *Lots of Science Library Book #4.*

Activities:

Observing Faults and Folds

Focus Skill: demonstrating concepts

Activity Materials: three different colors of clay or dough

Activity: Form each color of clay into the same size rectangle, about 1 inch thick. Place one rectangle on top of another. Place the layers of clay on a tabletop. Push in on each side until the middle moves up. This demonstrates an upward fold in the lithosphere. Try to make a downward fold with the clay.

Cut the clay into two pieces and demonstrate the three types of faults that occur in the lithosphere.

✎✎✎ After you complete the *Faults and Folds Graphic Organizer,* below, add sketches of this activity under the appropriate tab.

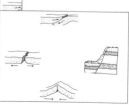

Faults and Folds – Graphic Organizer

Focus Skills: explaining data, labeling sketches

Paper Handouts: a copy of Graphic 4A   5" x 7 1/2" sheet of paper

Graphic Organizer: This is the beginning of a 22-Tab Graphic Organizer entitled *Landforms and Surface Features of Earth* that will be used in this and future lessons. Title the 5" x 7 1/2" sheet to make a cover. Glue the cover on the glue line of Graphic 4A. On Graphic 4A:

✎ Explain what you have learned about faults and folds. Color the illustrations. Draw your own examples on the left page.

✎✎ Use the *Lots of Science Library Book #4* to label the illustrations. Write clue words about faults and folds: *split in rock layers, shifting of land, bending of layers, crumpled land.*

✐✐✐ Complete ✎✎. Research the *Fascinating Physical Features of Earth* examples from the *Lots of Science Library Book #4* or other examples of faults and folds. Write a descriptive or expository paragraph about them on the left page.

Fascinating Facts about Earth – Graphic Organizer

Focus Skill: map reading
Paper Handouts: a copy of Graphic 4B   *Earth Shutter Fold Project*
Graphic Organizer: Cut out Graphic 4B and fold on the middle
    line so that the illustration is on the cover of the little
    book. Follow the directions below for the inside and glue
    it to the appropriate place on the world map inside the
    Earth Shutter Fold Project.

✎ Draw the cover picture on the inside and color it.
✎✎ Copy information from the *Lots of Science Library Book*
    about the cover picture: *San Andreas Fault is a 700 mile long crack.*
✎✎✎ Write information about the cover picture.

Experiences, Investigations, and Research

Select one or more of the following activities for individual or group enrichment projects. Allow your students to determine the format in which they would like to report, share, or graphically present what they have discovered. This should be a creative investigation that utilizes your students' strengths.

1. Research the San Andreas Fault and locate it on a map. This fault enters northern California from the Pacific Ocean and extends southeastward into the southern part of the state. When Earth's crust moves along this fault line, earthquakes are felt in California. Make a time line of movement along this fault line.

2. Draw a diagram of a cross-section of a folded mountain. Label the parts and write a caption explaining how your mountain was formed.

3. Research the formation of the Appalachian Mountains in the eastern United States. How did these folded mountains form? How has weathering and erosion affected them? What part did they play in the early history of America? Compare and contrast them to the Rocky Mountains in the western United States.

4. Make a Venn diagram showing what the Appalachian Mountains (U.S.A.) and the Alps (Europe) have in common. Note that both are examples of folded mountains. What else do they have in common?

5. http://www.educ.uvic.ca/Faculty/jtinney/earth%20science/ESMain.html

What are earthquakes?

Lithosphere Concepts:

- Any tremor or vibration in Earth's crust is called an earthquake.
- Earthquakes are most likely to occur at continental plate boundaries and fault lines.
- When continental plates suddenly shift, serious earthquakes result.
- The point in the lithosphere where the slip occurs is called the focus of the earthquake. The epicenter is the point on earth's surface above the focus.
- When the continental plates stop moving, rocks resettle, causing aftershocks in the area of the earthquake.
- Scientists who study earthquakes use two different scales to measure them, the Richter scale and the Mercalli scale.

Vocabulary: earthquake   focus   vibrate   aftershocks   *Richter scale
*Mercalli scale

Read: *Lots of Science Library Book #5.*

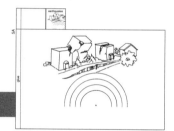

Activities:

Earthquakes – Graphic Organizer

Focus Skills: explaining concepts   labeling sketches
Paper Handouts: a copy of Graphic 5A
    Landforms and Surface Features of Earth
Graphic Organizer: Glue Graphic 5A under the previous page in *Landforms and Surface Features of Earth* at the glue line.

✎ Explain what you have learned about earthquakes. Color the illustrations. Draw your own examples on the left page.

✎✎ Use the *Lots of Science Library Book #5* to label the illustrations. Write clue words about earthquakes: *vibrations, slide in plates, focus, aftershocks.*

✎✎✎ Complete ✎✎. Research the *Fascinating Physical Features of Earth* examples from the *Lots of Science Library Book #5* or other examples of earthquakes. Write a descriptive or expository paragraph about them on the left page.

Fascinating Facts about Earth – Graphic Organizer

Focus Skill: map reading
Paper Handouts: a copy of Graphics 5B-C   *Earth Shutter Fold Project*
Graphic Organizer: Cut out Graphics 5B-C and fold on the middle line so that the illustration is on the cover of each little book. Follow the directions below for the inside and glue it to the appropriate place on the world map inside the *Earth Shutter Fold Project*.

✐   Draw the cover pictures on the inside and color them.

✐✐   Copy information from the *Lots of Science Library Book* about the cover pictures: *San Francisco earthquake, 1989, 7.1; Libon, Portugal earthquake, 1755, 10 minutes.*

✐✐✐   Write information about the cover pictures.

Experiences, Investigations, and Research

Select one or more of the following activities for individual or group enrichment projects. Allow your students to determine the format in which they would like to report, share, or graphically present what they have discovered. This should be a creative investigation that utilizes your students' strengths.

 1.   Using the clay from Lesson 4, show how an earthquake begins in the lithosphere.

2.   Compare and contrast the impact of earthquakes with shallow and deep focus points.

3.   Research the history of earthquake detection devices.

4.   Invent and build your own tool for measuring the seismic waves of an earthquake. Use something that vibrates, like a clothes dryer, to test your invention.

5.   Write a story about a person's life before, during, and after an earthquake.

6.   Investigate construction materials and techniques that make buildings safer during earthquakes. Research construction practices in countries such as Japan, where earthquakes are common.

7.   Research, diagram, and explain P&S waves.

8.   Compare and contrast the Richter Scale and the Mercalli Scale.

9.   Make a timeline of historic earthquakes.

10.   http://howstuffworks.lycoszone.com/earthquake.htm

11.   http://www.crustal.ucsb.edu/ics/understanding

How are mountains formed?

Lithosphere Concepts:

- A mountain is any point on Earth's surface that rises 1,000 feet or more above its surroundings. Hills are less than 1,000 feet in height.
- As Earth's continental plates move, the layers of the lithosphere can bend or crumple, creating folded mountains.
- As Earth's continental plates move, the layers of the lithosphere can break and slip along fault lines, creating block mountains.
- Sometimes molten rock, or magma, pushes Earth's surface upwards but does not break through. This creates dome mountains.
- When molten rock, or magma, does break through the surface of Earth, volcanic mountains are formed.

Vocabulary: mountain   folded   block   dome   volcanic

Read: *Lots of Science Library Book #6.*

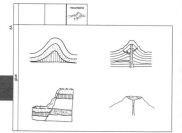

Activities:

Mountains – Graphic Organizer

Focus Skills: explaining concepts, labeling sketches
Paper Handouts: a copy of Graphic 6A
    Landforms and Surface Features of Earth made in previous lesson
Graphic Organizer: Glue Graphic 6A behind the previous page of *Landforms and Surface Features of Earth* on the glue line.

✎   Explain what you have learned about mountains.

✎✎   Use the *Lots of Science Library Book #6* to label the illustrations. Write clue words about mountains: *folded, block, dome, volcanic.*

✎✎✎   Complete ✎✎. Research the *Fascinating Physical Features of Earth* examples from the *Lots of Science Library Book #6* or other examples of mountains. Write a descriptive or expository paragraph about them on the left page/

Fascinating Facts about Earth – Graphic Organizer

Focus Skill: map reading
Paper Handouts: a copy of Graphics 6B-C   *Earth Shutter Fold Project*
Graphic Organizer: Cut out Graphics 6B-C and fold on the middle line so that the illustration is on the cover of each little book. Follow the directions below for the inside and glue it to the appropriate place on the world map inside the *Earth Shutter Fold Project.*

✐ Draw the cover pictures on the inside and color them.

✐✐ Copy information from the *Lots of Science Library Book* about the cover pictures: *Mt. Everest is 29,025 feet high; Andes Mountains are 4,496 miles long.*

✐✐✐ Write information about the cover pictures.

Experiences, Investigations, and Research

Select one or more of the following activities for individual or group enrichment projects. Allow your students to determine the format in which they would like to report, share, or graphically present what they have discovered. This should be a creative investigation that utilizes your students' strengths.

1. Using layers of clay, demonstrate the formation of the four types of mountains.

2. Locate the major mountain ranges of each continent on a map or globe. Notice that the mountain ranges of North and South America tend to run north and south, while the ranges of Europe and Asia tend to run east and west. The Ural Mountains, the one mountain range on the Eurasian continent that runs north and south, divides Europe from Asia.

3. Explain why some mountainous regions have frequent earthquakes.

4. Describe ways in which to determine the age of mountains. How are the Appalachian Mountains known to be older than the Rocky Mountains?

5. Research and describe the Mid-Atlantic Ridge as the world's largest mountain chain.

6. Compare and contrast mountains on continental land and mountains on the ocean floor.

7. Explain what happens when mountains rise above the ocean's surface.

8. Research the formation of the island of Surtsey.

9. http://www.uen.org/utahlink/lp_res/TRB023.html

What is a volcano?

Lithosphere Concepts:

- Volcanoes form when pressure builds below Earth's surface and molten rock, or magma, slowly rises along a crack or fissure, melting surrounding rock.
- Magma may cool and become solid in the Earth's crust, or it may break through the surface, at which time it is called lava.
- A volcano's shape is determined by the type of lava produced, how far it flows, and how forcefully it erupts.
- Some eruptions are so violent that all or part of the volcano blows away, leaving a crater.
- There are four classes of volcanoes:
    Active volcanoes are constantly erupting.
    Intermittent volcanoes erupt at regular intervals.
    Dormant volcanoes are inactive, but can become active again.
    Extinct volcanoes have been inactive for hundreds of years.

Vocabulary: volcano   magma   lava   vent   *eruption   *dormant   *extinct

Read: *Lots of Science Library Book #7.*

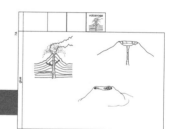

Activities:

Volcanoes – Graphic Organizer

Focus Skills: explaining concepts, labeling sketches
Paper Handouts: a copy of Graphic 7A
    Landforms and Surface Features of Earth
Graphic Organizer: Glue Graphic 7A under the previous page of *Landforms and Surface Features of Earth* on the glue line.

✎ Explain what you have learned about volcanoes. Color the illustrations. Draw your own examples on the left page.

✎✎ Use the *Lots of Science Library Book #7* to label the illustrations. Write clue words about volcanoes: *magma, lava, vent, eruption.*

✎✎✎ Complete ✎✎. Research the *Fascinating Physical Features of Earth* examples from the *Lots of Science Library Book #7* or other examples of volcanoes. Write a descriptive or expository paragraph about them on the left page.

Fascinating Facts about Earth – Graphic Organizer

Focus Skill: map reading
Paper Handouts: a copy of Graphics 7B-C   *Earth Shutter Fold Project*
Graphic Organizer: Cut out Graphics 7B-C and fold on the middle line so that the illustration is

on the cover of each little book. Follow the directions below for the inside and glue it to the appropriate place on the world map inside the *Earth Shutter Fold Project*. With a red pen mark the "ring of fire" on the map.

✎ Draw the cover pictures on the inside and color them.

✎✎ Copy information from the *Lots of Science Library Book* about the cover pictures.

✎✎✎ Write information about the cover pictures.

7B   7C

Make a Volcano

Focus Skill: following directions

Activity Materials: vinegar   baking soda   bottle or jar   paper towel
optional: red food coloring

Activity: Follow the directions below to make an eruption. If desired, you can bury the bottle or jar in a mountain of dirt, being sure to leave the top open.

1. Fill the bottle 1/3 full of vinegar. Add food coloring, if desired.
2. Place the bottle outside on a smooth surface.
3. Put 2 tablespoons of baking soda on the paper towel.
4. Roll the towel loosely, pinching the ends.
5. Drop the paper towel package in the vinegar and stand back.

Teacher's Note: As the paper towel opens, the baking soda will react with the vinegar and erupt.

Experiences, Investigations, and Research

Select one or more of the following activities for individual or group enrichment projects. Allow your students to determine the format in which they would like to report, share, or graphically present what they have discovered. This should be a creative investigation that utilizes your students' strengths.

✳ 📚 1. Research one of the following historic volcanic eruptions: Pompeii, Fugi, Vesuvilus, Mt. Saint Helen, or Kilaminjaro.

∩ 👐 2. Using small sheets of stiff paper, draw a volcano erupting in sequence. First draw a volcano that is not erupting. Next draw a volcano with only a little lava being emitted. In the 8 remaining pictures, draw a little more ash, lava, and gas blowing out of the volcano, until the 10th picture shows a huge volcanic eruption. Place the pictures in order from no eruption to a full eruption. Staple them together along the left side. Flip the pages to see your volcano erupt.

∩ 📚 3. Research seamounts or volcanoes that form on the ocean floor. Compare and contrast two types of seamounts – a guyot and a volcanic island. Investigate and report on the newest Hawaiian Island formation, Loihi Seamount.

✓ 💻 4. http://visibleearth.nasa.gov/Solid_Earth/Volcanoes/

✓ 💻 5. http://pubs.usgs.gov/gip/volc/nature.html

What are tsunamis, hot springs, geysers, and volcanic islands?

Lithosphere Concepts:

- Shock waves of a large underwater earthquake or volcano can cause giant walls of water. These are called tsunamis.
- Tsunamis are not very noticeable at sea, but by the time they reach land they can be 6 – 60 feet (1.8 - 18m) high traveling at 500 – 600 mph (805 - 965 kmph).
- In areas of volcanic activity, hot zones heat underground water.
- Heated water can bubble out of cracks forming hot springs.
- Underground water can become trapped. As it heats, the water expands and pressure builds until the water bursts through the surface as a geyser.
- Volcanic islands are formed from underwater volcanoes.

Teacher's Note: An alternative assessment suggestion for this lesson is found on pages 64-65. If Graphics Pages are being consumed, first photocopy any assessment graphics that will be needed.

Vocabulary: hot springs   geyser   volcanic

*tsunamis

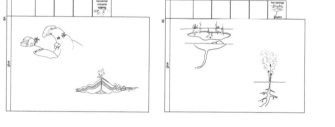

Read: *Lots of Science Library Book #8.*

Activities:

Tsunamis, Hot Springs, Geysers, and Volcanic Islands –

Focus Skills: explaining processes   communicating information
Paper Handouts: a copy of Graphics 8A-B   *Landforms and Surface Features of Earth*
Graphic Organizer: Glue Graphic 8A under the previous page of *Landforms and Surface Features of Earth,* on the glue line. Glue Graphic 8B on the glue line under Graphic 8A in *Landforms and Surface Features of Earth.*

✎ Explain what you have learned about tsunamis, hot springs, geysers, and volcanic islands. Color the illustrations. Draw your own examples on the left page.

✎✎ Use the *Lots of Science Library Book #8* to label the illustrations. Write clue words about each feature: tsunamis: *earthquake or volcano underwater, high waves;* hot springs: *heated water, comes up in cracks;* geysers: *heated water with pressure, shoots up;* volcanic islands: *underwater volcano, lava cools, grows above ocean level.*

✎✎✎ Complete ✎✎. Research the *Fascinating Physical Features of Earth* examples from the *Lots of Science Library Book #8* or other examples of these features. Write a descriptive or expository paragraph about them on the left page.

Focus Skill: map reading
Paper Handouts: a copy of Graphics 8C-F   *Earth Shutter Fold Project*
Graphic Organizer: Cut out Graphics 8C-F and fold on the middle line so that the illustration is
on the cover of each little book. Follow the directions below for the inside and glue it
to the appropriate place on the world map inside the
Earth Shutter Fold Project.

✎   Draw the cover pictures on the inside and color them.
✎✎   Copy information from the *Lots of Science Library Book* about the cover pictures.
✎✎✎ Write information abut the cover pictures.

Experiences, Investigations, and Research

Select one or more of the following activities for individual or group enrichment projects. Allow
your students to determine the format in which they would like to report, share, or graphically
present what they have discovered. This should be a creative investigation that utilizes your
students' strengths.

1. Investigate the history of the region now known as Yellowstone National Park from
    prehistoric times to the present.

2. Research hydrothermal vents and explain why some of them form chimney-like
    structures that are made of mineral deposits rich in iron, copper, and zinc. Describe
    how life has adapted to the temperature extremes in these areas, and how it is
    based upon chemosynthesis.

3. Draw and label the following: fumaroles, geysers, and bubbling mud called mud
    pots or mud volcanoes.

4. Research volcanic islands such as the Aleutian Islands, the islands of Japan, or the
    Lesser Antilles.

5. Draw a map of the Hawaiian Islands. Explain how and why a new island is
    forming.

6. http://www.geophys.washington.edu/tsunami/general/physics/physics.html

7. http://pubs.usgs.gov/gip/volc/geysers.html

Notes

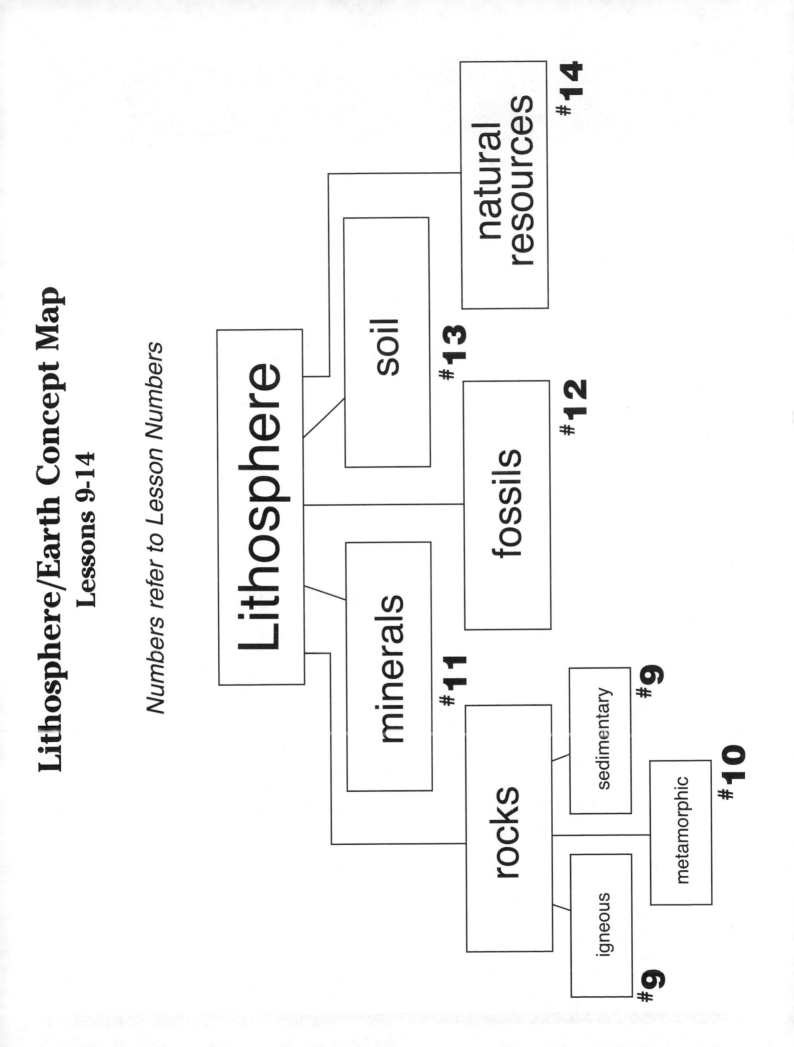

Lithosphere/Earth Concept Map
Lessons 9-14

Numbers refer to Lesson Numbers

Lithosphere

soil #13

minerals #11

fossils #12

natural resources #14

rocks

sedimentary #9

igneous #9

metamorphic #10

What are igneous and sedimentary rocks?

Lithosphere Concepts:

- There are three main types of rock: igneous, sedimentary, and metamorphic.
- Igneous rock is formed when magma rises from Earth's mantle, cools, and becomes solid.
- Sedimentary rock is formed from the deposition of mineral and organic sediments.
- Different types of sedimentary rocks are formed depending upon the sediment deposited and the manner in which it is compacted and cemented.

Teacher's Note: For an in-depth study of rocks and minerals, see *Great Science Adventures, Discovering Rocks and Minerals.*

Vocabulary: igneous   sedimentary   *organic   *inorganic

Read: *Lots of Science Library Book #9.*

Activities:

Types of Rocks – Graphic Organizer

Focus Skills: describing a category   comparing and contrasting
Paper Handouts: a copy of Graphic 9A   *Landforms and Surface Features of Earth*
Graphic Organizer: Glue Graphic 9A on the glue line under the previous page in *Landforms and Surface Features of Earth.*

✎   Explain what you have learned about igneous and sedimentary rocks. Color the illustrations. Draw your own examples on the left page.

✎✎   Use the *Lots of Science Library Book #9* to label the illustrations. Write clue words about each type of rock: igneous: *cooled magma;* sedimentary: *layers of sediment.*

✎✎✎   Complete ✎✎. Research igneous and sedimentary rocks. Write a descriptive or expository paragraph about them on the left page. Compare and contrast the formation of igneous and sedimentary rock.

Teacher's Note: Store this book for use in Lesson 10.

Experiences, Investigations, and Research

Select one or more of the following activities for individual or group enrichment projects. Allow your students to determine the format in which they would like to report, share, or graphically present what they have discovered. This should be a creative investigation that utilizes your students' strengths.

  1. Begin collecting rocks. Sort the rocks as igneous or sedimentary. What clues will you look for? Use rock and mineral identification manuals to help you name your specimens. What percentage of the rocks collected were sedimentary? igneous? unidentifiable?

2. Design a demonstration to show how sediments might be deposited in layers. For example, find two different kinds of soil, light-colored sand and dark soil. Place 1/2 cup of each in a jar, fill the jar half full of water, and shake well. Watch how the particles settle out of the water and form layers of sediment.

3. Look for pumice stone in the grocery store or hardware store. Examine its porous structure. Read to discover what it is used for.

4. Note buildings made out of igneous and sedimentary rocks. Compare and contrast these two building materials. Sometimes fossils can be seen on the sides of buildings or in sidewalks made out of sedimentary rock. You can go on a fossil hunt in a city!

5. http://www.fi.edu/fellows/payton/rocks/create/index.html

Great Science Adventures

Lesson 10

What are metamorphic rocks?

Lithosphere Concepts:

- Metamorphic rock is formed when heat, pressure, or both alter igneous, sedimentary, or other metamorphic rocks.
- The rock cycle shows how rocks can change and be transformed.

Vocabulary: metamorphic   rock cycle

Read: *Lots of Science Library Book #10.*

Activities:

Types of Rocks – Graphic Organizer

Focus Skills: describing a category
          comparing and contrasting

Paper Handouts: a copy of Graphic 10A   *Landforms and Surface Features of Earth*

Graphic Organizer: Glue Graphic 10A on the glue line, under the previous page of *Landforms and Surface Features of Earth.*

✎ Explain what you have learned about metamorphic rocks. Color the illustrations. Draw your own examples on the left page.

✎✎ Use the *Lots of Library Book #10* to label the illustrations. Write clue words about metamorphic rocks: *heat and pressure changes rock.*

✎✎✎ Complete ✎✎. Research metamorphic rocks or the Rock Cycle. Write a descriptive or expository paragraph about them on the left page.

Experiences, Investigations, and Research

Select one or more of the following activities for individual or group enrichment projects. Allow your students to determine the format in which they would like to report, share, or graphically present what they have discovered. This should be a creative investigation that utilizes your students' strengths.

 1. Add to your rock collection. Try to determine if you have metamorphic rocks in your collection. Share your collection with a friend.

2. Investigate the following metamorphic rocks: marble, schist, slate.

3. Explain the rock cycle in your own words.

 4. http://www.fi.edu/fellows/payton/rocks/create/index.html

Notes

What are minerals?

Lithosphere Concepts:

- Minerals are naturally-made, nonliving, crystalline substances made of elements.
- Scientists have identified nearly 3,000 different minerals in the lithosphere.
- Most minerals develop in liquids such as magma. Magma contains all the kinds of atoms that make up the earth's minerals. As magma cools, crystals of different shapes form, resulting in different kinds of minerals.
- Minerals are graded by their hardness using the Mohs scale.
- Minerals can be divided into two groups: native elements and compound minerals.
- Native elements are minerals made up of a pure element.
- Compound minerals are combinations of two or more elements.

Vocabulary: minerals   elements   *Mohs scale   *native element   *compound element

Read: *Lots of Science Library Book #11.*

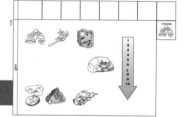

Activities:

Minerals in Earth – Graphic Organizer

Focus Skills: describing substances

Paper Handouts: a copy of Graphic 11A   *Landforms and Surface Features of Earth*

Graphic Organizer: Glue Graphic 11A on the glue line under the previous page of *Landforms and Surface Features of Earth.*

✏ Explain what you have learned about minerals. Color the illustrations. Draw your own examples on the left page.

✏✏ Use the *Lots of Science Library Book #11* to label the illustrations. Write clue words about minerals: *naturally-made, nonliving substances, more than 3,000 different minerals, graded by their hardness, native elements and compound minerals.*

✏✏✏ Complete ✏✏. Research the *Fascinating Physical Features of Earth* examples from the *Lots of Science Library Book #11* or other examples of mineral deposits. Write a descriptive or expository paragraph about them on the left page.

Fascinating Facts about Earth – Graphic Organizer

Focus Skill: map reading

Paper Handouts: a copy of Graphic 11B   *Earth Shutter Fold Project*

Graphic Organizer: Cut out Graphic 11B and fold on the middle line so that the illustration is on the cover of the little book. Follow the directions below for the inside and glue it to the appropriate place on the world map inside the *Earth Shutter Fold Project.*

11B

✎ Draw the cover picture on the inside and color it.

✎✎ Copy information from the *Lots of Science Library Book* about the cover picture.

✎✎✎ Write information about the cover pictures.

Experiences, Investigations, and Research

Select one or more of the following activities for individual or group enrichment projects. Allow your students to determine the format in which they would like to report, share, or graphically present what they have discovered. This should be a creative investigation that utilizes your students' strengths.

 1. Make a Venn diagram comparing and contrasting diamonds and graphite polymorphic minerals made from the element carbon.

✓ 2. Make a table to record information on the composition and structure of the major classes of minerals: elements, sulfides, halides, carbonates, sulfates, oxides, phosphates, and silicates.

3. Make up a guessing game to play with your friends. For example, think of something and then give your friends clues to see who can guess it first. For the first clue you might tell them you are you thinking of something that is a plant, an animal, or a mineral.

4. Investigate bottled mineral water. What is the history of its use and popularity?

✓ 5. What minerals does the human body need to stay healthy? What minerals do plants need? How do living organisms obtain needed minerals? What are the results of mineral deficiencies?

✓ 6. http://www.trickyricky.addr.com/minerals.htm

What are fossils?

Lithosphere Concepts:

- Fossils are the preserved remains of organisms that were once living.
- For an organism to become a fossil, it must be rapidly buried in sediment, or something similar, and not destroyed by Earth's heat, pressure, or weathering.
- The hard body parts of animals - bones, teeth, claws, shells, scales, and tusks - are commonly found as fossils.
- Plant fossils include leaves, seeds, and petrified wood.
- Mold fossils are created when an organism decays but leaves an impression in the rock.
- Cast fossils are the impression of an organism that has been filled in with minerals and sediment. The original organism is gone, but a mold exists of its form.

Vocabulary: fossil   decay   *fossilization   *petrified

Read: *Lots of Science Library Book #12.*

Activities:

Fossils in Earth – Graphic Organizer

Focus Skill: describing substances
Paper Handouts: a copy of Graphic 12A   *Landforms and Surface Features of Earth*
Graphic Organizer: Glue Graphic 12A behind the previous page in the *Landforms and Surface Features of Earth.*

✎   Explain what you have learned about fossils. Color the illustrations. Draw your own examples on the left page.

✎✎   Use the *Lots of Science Library Book #12* to label the illustrations. Write clue words about fossils: *preserved remains of plants and animals, hard parts of an animal or can be an impression in rock, either filled with sediment or not.*

✎✎✎   Complete ✎✎. Research fossils. Write a descriptive or expository paragraph about them on the left page.

Impressions

Focus Skill: demonstrate a concept
Activity Materials: dirt   plaster of Paris   water
Activity: Put the dirt in a pan or use it outside. Be sure the dirt is moist enough to hold an impression. Make an impression of your hand or foot in the dirt. Mix the plaster of Paris with water and pour it into the impression. When it is dry, lift the impression out.

Teacher's Note: This is a demonstration that gives the students an idea of how some fossils are made; it does not create a fossil.

Experiences, Investigations, and Research

Select one or more of the following activities for individual or group enrichment projects. Allow your students to determine the format in which they would like to report, share, or graphically present what they have discovered. This should be a creative investigation that utilizes your students' strengths.

 1. Press a shell into a piece of clay. Remove the shell. This is what a mold fossil looks like. Cover the clay imprint with oil. Mix some dirt and water to make mud. Fill the imprint with the liquid mud. Allow it to dry until hard. Carefully remove the clay to expose a cast fossil.

 2. Research trace fossils: footprints and track ways, trails and burrows, eggs and nests, petrified feces or skin imprints.

 3. Investigate the fossils found in your area. What fossils are common? Why?

 4. Start a fossil collection. Search the Internet for information on fossils and fossil identification aids.

5. http://www.don-lindsay-archive.org/creation/fossil_def.html

What is soil?

Lithosphere Concepts:

- Soil is the layer of weathered rocks and minerals that forms on the surface of Earth's crust. Its depth varies from less than an inch in thickness to hundreds of feet thick.
- Soil is made up of weathered rocks, water, air, and decaying organic material.
- Five main factors affect the formation and type of soil: weathered parent rock, organic matter, climate, time, and the geography of the land.

Vocabulary: soil   climate   geography

Read: *Lots of Science Library Book #13.*

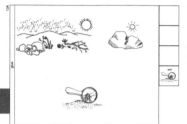

Activities:

Earth's Soil – Graphic Organizer

Focus Skill: describing substances
Paper Handouts: *Landforms and Surface Features of Earth*   a copy of Graphic 13A
Graphic Organizer: Glue Graphic 13A under the previous page in *Landforms and Surface Features of Earth.*

✎ Explain what you have learned about soil. Color the illustrations. Draw your own examples on the left page.

✎✎ Use the *Lots of Science Library Book #13* to label the illustrations. Write clue words about soil: *loose material that covers land, weathered rocks, water, air, rotting organic material,* and *geography.*

✎✎✎ Complete ✎✎. Research the *Fascinating Physical Features of Earth* examples from the *Lots of Science Library Book #13* or other examples of soil use. Write a descriptive or expository paragraph about them on the left page.

Air Space in Soil – Investigative Loop – Lab 13-1

Lab 13-1

Focus Skill: calculating with a formula
Lab Materials: soil   2 measuring cups   water
Paper Handouts: 8.5" x 11" sheet of paper   a copy of Lab Graphic 13-1
          Lab Record Cards (index cards or 3"x4" pieces of paper)
Graphic Organizer: Make a Pocket Book. Glue Lab Graphic 13-1 on the left pocket. This is the Lab Book. It will be used in this and future lessons.
Question: How much air is in soil?
Research: Read *Lots of Science Library Book #13* and review the Question.
Procedure: Put soil in the measuring cup. Shake it until it is level. Slowly add the same amount of water to the measuring cup. For example, add 5 ounces of water to 5 ounces of soil. Allow the water and soil to settle.

Observations: Observe the amount of water and soil in the measuring cup after it has settled.

Record the Data: Label a Lab Record Card "Lab 13-1" and draw the lab or explain the Procedure of the lab. Record the amount of soil put into the measuring cup. Record the amount of water put into the measuring cup. Read and record the amount in the measuring cup after the mixture has settled.

Conclusions: The difference between the amount of soil added to the amount of water put into the measuring cup and how the mixture measures when it settles indicates the amount of air in the soil. In our example, if 5 ounces of soil and 5 ounces of water are put into the measuring cup, but the mixture measures 8 ounces, then 2 ounces of air were in the soil (5 + 5 = 10 10 − 8 = 2). Use this formula to determine how much air was in the soil used in this lab.

Communicate the Conclusions: Label a Lab Record Card "Lab 13-1" and write the conclusions about the amount of air in the soil.

Spark Questions: Discuss questions sparked by this lab.

New Loop: Choose one question to investigate further, or create an *Investigative Loop* using several different samples of soil.

✎✎✎ **Design Your Own Experiment:** Select a topic based upon this *Investigative Loop* experience. See page vii for more details.

Organic Materials in Soil – Investigative Loop – Lab 13-2

Focus Skill: predicting outcome

Lab Materials: 3 different soil samples   3 jars   water

Paper Handouts: a copy of Lab Graphic 13-2   Lab Book   Lab Record Cards

Graphic Organizer: Glue Lab Graphic 13-2 on the right pocket of the Lab Book.

Lab 13-2

Question: What soil sample has the most organic material in it?

Research: Look through all three soil samples. Read *Lots of Science Library Books #13* and review the Question.

Prediction: Find a manner of labeling each soil sample, such as putting each one in a plastic bag and marking the bags A, B, and C. Label a Lab Record Card "Lab 13-2" and write your prediction on the card.

Procedure: Put the same amount of soil in each jar. Label the containers as you did the soil. Put the same amount of water in each jar, and shake or stir each jar vigorously for a minute. Place the jars in a secure place for 24 hours.

Observations: Observe the floating material in each jar. It is the organic material that was in each soil sample.

Record the Data: Draw or describe the lab on a Lab Record Card that has been labeled "Lab 13-2." Record the amount of soil and water in each jar. Describe the amount of floating organic material observed in each jar. Use qualitative observations such as "a little bit of organic material," "more organic material than soil B," or "the most organic material."

Conclusions: Based on the data collected, which soil sample had the most organic material in it? Compare this to the prediction you made about this lab.

Communicating the Conclusions: Label a Lab Record Card "Lab 13-2" and write the conclusions about this lab. Include sketches, if applicable.

Spark Questions: Discuss questions sparked by this lab.

New Loop: Choose one question to investigate further.

✎✎✎ **Design Your Own Experiment:** Select a topic based upon this *Investigative Loop* experience. See page vii for more details.

Focus Skill: map reading
Paper Handouts: a copy of Graphics 13B-C   *Earth Shutter Fold Project*
Graphic Organizer: Cut out Graphics 13B-C and fold on the middle line so that the illustration
  is on the cover of each little book. Follow the directions
  below for the inside and glue it to the appropriate place
  on the world map inside the *Earth Shutter Fold Project*.

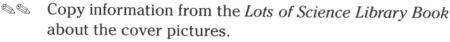

  ✎ Draw the cover pictures on the inside and color them.
  ✎✎ Copy information from the *Lots of Science Library Book*
      about the cover pictures.
  ✎✎✎ Write information about the cover pictures.

Experiences, Investigations, and Research

Select one or more of the following activities for individual or group enrichment projects. Allow
your students to determine the format in which they would like to report, share, or graphically
present what they have discovered. This should be a creative investigation that utilizes your
students' strengths.

 1. Investigate soil by using a filter and a colander to separate it into parts by size and
       composition. Describe what you find.

 2. Explain why and how soil contents change. Describe why and how gardeners
       change soil content.

3. Compare and contrast the soil found on the floor of a rainforest to the soil found in
    grasslands or in someone's garden. Which would be more fertile? Why?

4. What is topsoil? How does it form? How long does it take to form? How can it be
    preserved?   What factors destroy or harm topsoil? Why is topsoil sold?

5. Ask permission to dig a hole. Observe the layers of soil that you uncover. Sketch a
    cross-section of the soil layers and write about each. Is the soil fine or rough in
    texture? What color is it? What do you see in the soil?

6. Research hydroponics, or the raising of plants in nutrient-rich water. Explain how
    water can take the place of soil.

7. http://www.epa.gov/gmpo/edresources/soil.html

Notes

What are Earth's mineral resources?

Lithosphere Concepts:

- Mineral resources are the natural substances taken out of Earth's lithosphere and used as fuels or raw materials. These include coal, oil, and gas.
- Fossil fuels are made from decayed plants and animals that have chemically changed over time due to pressure and heat from the lithosphere.

Teacher's Note: An alternative assessment suggestion for this lesson is found on pages 64-65. If Graphics Pages are being consumed, first photocopy the assessment graphics that are needed.

Vocabulary: resources   fuel   coal   oil   gas   *chemically   *peat   *lignite

Read: *Lots of Science Library Book #14.*

Activities:

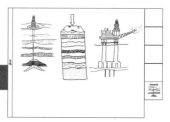

Earth's Fossil Fuels – Graphic Organizer

Focus Skill: describing substances
Paper Handouts: *Landforms and Surface Features of Earth*
   a copy of Graphic 14A
Graphic Organizer: Glue Graphic 14A under the previous page in the *Landforms and Surface Features of Earth.*

✎ Explain what you have learned about Earth's mineral resources. Color the illustrations. Draw your own examples on the left page.

✎✎ Use the *Lots of Science Library Book #14* to label the illustrations. Write clue words about Earth's mineral resources: *coal, oil, and gas; plant and animal materials chemically changed by pressure and heat.*

✎✎✎ Complete ✎✎. Research the *Fascinating Physical Features of Earth* examples from the *Lots of Science Library Book #14.* Write a descriptive or expository paragraph about them on the left page.

Fascinating Facts about Earth – Graphic Organizer

Focus Skill: map reading
Paper Handouts: a copy of Graphics 14B-G   *Earth Shutter Fold Project*
Graphic Organizer: Cut out Graphics 14B-G and fold on the middle line so that the illustration is on the cover of each little book. Follow the directions below for the inside and glue it to the appropriate place on the world map inside the *Earth Shutter Fold Project.*

14F   14E
14G   14D
14B
14C

✎ Draw the cover pictures on the inside and color them.

✎✎ Copy information from the *Lots of Science Library Book* about the cover pictures.

✎✎✎ Write information about the cover pictures.

Experiences, Investigations, and Research

Select one or more of the following activities for individual or group enrichment projects. Allow your students to determine the format in which they would like to report, share, or graphically present what they have discovered. This should be a creative investigation that utilizes your students' strengths.

 1. Fossil Fuels
   http://www.fe.doe.gov/education/

Notes

Lithosphere/Earth Concept Map

Lessons 15-24

Numbers refer to Lesson Numbers

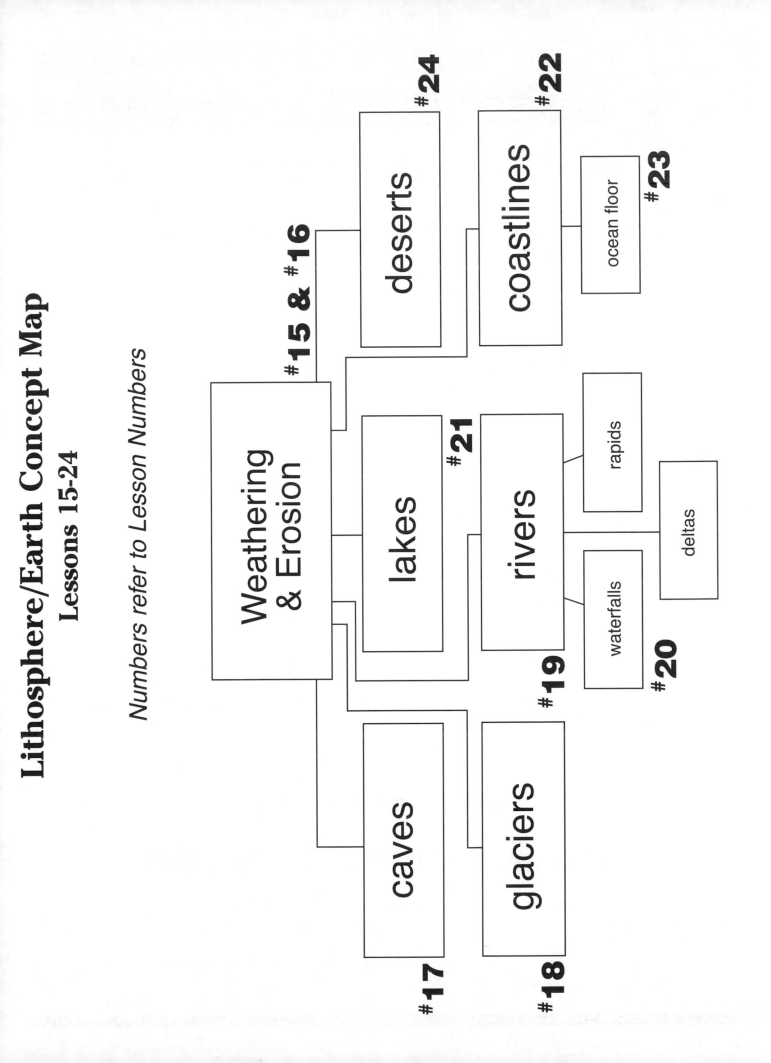

What is weathering and erosion?

Lithosphere Concepts:

- When rocks are exposed to rain, frost, and wind, they gradually break down. This weather related process is called weathering.
- There are two types of weathering - physical and chemical.
- Examples of physical weathering include the effects of rain, wind, temperature, and human influences.
- Chemical weathering occurs when chemicals wear away the minerals in rocks, or when acidic water dissolves the minerals in rocks.
- Pieces of rock broken off a parent rock are called debris. Debris can be carried by water, ice, and wind. As debris is carried, it grinds against rocks and land, causing them to wear down. This process is called erosion.

Vocabulary: weathering   physical   chemical   erosion   debris

Read: *Lots of Science Library Book #15.*

Activities:

Weathering – Graphic Organizer

Focus Skills: describing concepts, explaining processes

Paper Handouts: *Landforms and Surface Features of Earth*   a copy of Graphic 15A

Graphic Organizer: Glue Graphic 15A under the previous page of *Landforms and Surface Features of Earth.*

   ✎ Explain what you have learned about weathering. Color the illustrations. Draw your own examples on the left page.

   ✎✎ Use the *Lots of Science Library Book #15* to label the illustrations. Write clue words about weathering: *physical: temperature changes, freeze-thaw, animals, plants; chemical: rain, wearing away minerals.*

   ✎✎✎ Complete ✎✎. Research physical and chemical weathering. Write a descriptive or expository paragraph about them on the left page.

Erosion – Graphic Organizer

Focus Skills: describing concepts, explaining processes, comparing and contrasting

Paper Handouts: *Landforms and Surface Features of Earth*   a copy of Graphic 15B

Graphic Organizer: Glue Graphic 15B under the previous page of *Landforms and Surface Features of Earth.*

   ✎ Explain what you have learned about erosion. Color the illustrations. Draw your own examples on the left page.

   ✎✎ Use the *Lots of Science Library Book #15* to label the illustrations. Write clue words about erosion: *debris is carried by water, ice, or wind, riverbeds and banks, wind erodes rocks, ice erodes land, and waves erode shores.*

✎✎✎ Complete ✎✎. Research physical and chemical weathering. Write a descriptive or expository paragraph abut them on the left page. Compare weathering and erosion. How are they alike? How do they differ?

Freeze-thaw Process – Investigative Loop – Lab 15-1

Lab 15-1

Focus Skills: predicting outcome, drawing conclusions

Lab Materials: modeling clay   water   plastic wrap   use of a freezer

Paper Handouts: 8.5" x 11" sheet of paper   a copy of Lab Graphic 15-1
   Lab Book   Lab Record Cards

Graphic Organizer: Make a Pocket Book and glue it side-by-side to the Lab Book. Glue Lab Graphic 15-1 on the left pocket.

Question: How does the freeze-thaw process affect clay?

Research: Read *Lots of Science Library Book #15* and review the Question.

Predictions: Predict how clay will react to being frozen, thawed out, and then frozen again. Label a Lab Record Card "Lab 15-1" and write your prediction on it.

Procedure: Divide the clay into two pieces. Roll each piece into a ball, moistening it with water. Wrap each ball in plastic wrap and label one A and the other B. Place one in the freezer and one on an indoor shelf. After 24 hours, remove the frozen clay ball from the freezer, let it thaw out, then moisten it with water again. Refreeze it for another 24 hours. Unwrap both balls of clay and examine them.

Observations: Qualitatively describe the balls of clay before either of them was frozen. Describe the clay balls at the end of the lab.

Record the Data: Label two Lab Record Cards "Lab 15-1." On one card, write A and describe that ball of clay at the beginning and end of the lab. On the other card, write B and describe that ball of clay at the beginning and end of the lab. Use sketches if possible.

Conclusions: Review the data for this lab. Draw conclusions from the descriptions of each ball of clay. Compare the conclusions with the predictions.

Communicate the Conclusions: Use the Lab Record Card where the Predictions were written and write the conclusions about the lab. Share this with one other person.

Spark Questions: Discuss questions sparked by this lab.

New Loop: Choose one question to investigate further.

✎✎✎ **Design Your Own Experiment:** Select a topic based upon this *Investigative Loop* experience. See page vii for more details.

Experiences, Investigations, and Research

Select one or more of the following activities for individual or group enrichment projects. Allow your students to determine the format in which they would like to report, share, or graphically present what they have discovered. This should be a creative investigation that utilizes your students' strengths.

 1. Compare and contrast weathering and erosion.

 2. Look for examples of weathering and erosion where you live. For example, watch for ditches that wash deeper with each rain and look for debris that has been carried by water or wind.

3. Investigate ways in which erosion can be stopped.

4. Research how people who farm on sloped land or on mountain sides battle erosion.

How does erosion affect Earth?

Lithosphere Concepts:

- Water moving in streams and rivers can carve deep canyons through the process of erosion.
- Glaciers can erode away the sedimentary rock in an area, leaving the granite bedrock exposed.
- Sometimes water erodes sedimentary rock leaving only the rock that is resistant to erosion. This can result in mountain like formations.
- A mesa is a high, broad, flat-topped hill, surrounded by eroded debris.
- A butte is similar to a mesa with a smaller, narrower top and steep sides.
- Entire mountain ranges can erode, leaving behind only hard rock formations called monadnocks.

Teacher's Note: An alternative assessment suggestion for this lesson is found on pages 64-65. If Graphics Pages are being consumed, first photocopy assessment graphics that are needed.

Vocabulary: canyon   mesa   butte   *monadnock

Read: *Lots of Science Library Book #16.*

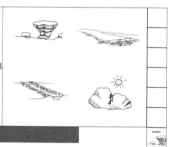

Activities:

Land Features created by Erosion – Graphic Organizer

Focus Skills: describing features, explaining processes, comparing and contrasting
Paper Handouts: *Landforms and Surface Features of Earth*
Graphic Organizer: On Graphic 15B used in Lesson 16:
- ✎ Color the illustrations. Draw your own examples on the left page.
- ✎✎ Use the *Lots of Science Library Book #16* to label the illustrations. Write clue words about erosion features: *canyons, granite core, mesa, butte.*
- ✎✎✎ Complete ✎✎. Research the *Fascinating Physical Features of Earth* examples from the *Lots of Science Library Book #16* or other examples of erosion features. Write a descriptive or expository paragraph about them on the left page.

Eroding Soil

Focus Skill: explaining a concept
Activity Materials: plastic bucket   large bowl   dirt   salt   water
Activity: Mix 1 cup of salt in water and add enough dirt to make it firm. Put the mixture in the bucket and turn the bucket over to make a dirt tower. Do the same with unsalted water

and dirt. Let the two dirt towers dry.

Discuss the Activity: What happened to each of the dirt towers? What was different about the towers? How do you explain what happened to the towers when they dried?

Note: The salt should help the first dirt tower not crumble. It acts like a cement with the dirt.

The Eroding Landscape

Note: This activity may be best completed outside.

Focus Skill: observing a concept

Activity Materials: 2 quarters   2 nickels   plate   dirt
          watering can

Activity: Place a mound of dirt on the plate. Put the coins at different locations on the dirt. Use the watering can to simulate rain on the plate.

Discuss the Activity: Describe the dirt mound after the rainfall. Why are some areas different than other areas? What land features do the elevated areas remind you of? **Possible answers: mesa or butte**

Fascinating Facts about Earth – Graphic Organizer

Focus Skill: map reading

Paper Handouts: a copy of Graphics 16A-B   *Earth Shutter Fold Project*

Graphic Organizer: Cut out Graphics 16A-B and fold on the middle line so that the illustration is on the cover of each little book. Follow the directions below for the inside and glue it to the appropriate place on the world map inside the *Earth Shutter Fold Project.*

    Draw the cover pictures on the inside and color them.

    Copy information from the *Lots of Science Library Book* about the cover pictures.

    Write information about the cover pictures.

Experiences, Investigations, and Research

Select one or more of the following activities for individual or group enrichment projects. Allow your students to determine the format in which they would like to report, share, or graphically present what they have discovered. This should be a creative investigation that utilizes your students' strengths.

1. Use clay to form a mesa, a butte, or both. Compare and contrast the two.

2. Write a story that takes place in the American Southwest on a mesa.

3. Research Stone Mountain, Georgia. Report on your findings.

4. Investigate hydrodynamics.

5. Describe erosion caused by rapids and waterfalls.

6. Compare and contrast erosion caused by moving surface water and moving underground water.

7. Erosion: http://www.marshfield.k12.wi.us/science/biology/eproject/erosion/erosion.htm

What are caves?

Lithosphere Concepts:

- Caves are formed when weaknesses in rock layers underground are eroded by water.
- Caves are most commonly found in limestone, although they can form in other types of rocks.
- As water moves through rock passageways, it dissolves minerals that are often left behind when the water evaporates.
- Stalactites and stalagmites form when ground water dissolves the mineral calcite (calcium carbonate) from limestone. The water enters the cave and evaporates, leaving calcite behind. Dripping water slowly deposits the calcite, forming stalactites and stalagmites.

Vocabulary: cave   limestone   structures   evaporate   pillar   *stalactite   *stalagmite

Read: *Lots of Science Library Book #17.*

Activities:

Caves –Graphic Organizer

Focus Skill: describing a process
Paper Handouts: a copy of Graphic 17A   *Landforms and Surface Features of Earth*
Graphic Organizer: Glue Graphic 17A under the previous page on the *Landforms and Surface Features of Earth.*

✎ Explain what you have learned about caves. Color the illustrations. Draw your own examples on the left page.

✎✎ Use the *Lots of Science Library Book #17* to label the illustrations. Write clue words about caves: *water seeps in cracks, erodes rock, water evaporates, leaves minerals, structures formed in caves.*

✎✎✎ Complete ✎✎. Research the *Fascinating Physical Features of Earth* examples from the *Lots of Science Book #17* or other examples of caves. Write a descriptive or expository paragraph about them on the left page.

Seeping Water – Investigative Loop – Lab 17-1

Focus Skills: quantitative observations   comparing and contrasting
Lab Materials: 3 soil samples (in plastic bags, labeled A, B, and C)   cheesecloth
plaster 2-liter bottle   rubber band   measuring cup   water
Paper Handouts: Lab Book   Lab Record Cards   a copy of Lab Graphic 17-1
Graphic Organizer: Glue Lab Graphic 17-1 on the right pocket.
Question: Which of the three soil samples will allow more water to seep through?

Research: Investigate each soil sample to determine your predictions about how easily it will allow water to seep through, or its permeability.

Predictions: Label a Lab Record Card "Lab 17-1" and write your predictions about the Question.

Procedure: Cut the top off the 2-liter bottle. Take the lid off and secure the cheesecloth on the small opening with the rubber band to make a funnel. Put soil sample A in the funnel. Place the plastic bottle top in the measuring cup, cheesecloth side down. Gradually pour a measured amount of water into the soil. Clean out the measuring cup and plastic bottle. Repeat the procedure for soil samples B and C. Be sure to pour in the same amount of water for each soil sample.

Lab 17-1

Observations: Note the time when the water is poured into each soil sample. Note the time when the water stops flowing into the measuring cup. Observe the amount of water in the measuring cup.

Record the Data: Label 3 Lab Record Cards "Lab 17-1" and write the letters A, B, and C on a card. On each card, list the time the water was poured into the soil, the time it quit flowing into the measuring cup, and the amount of water that was in the measuring cup.

Conclusions: Review the data. Which soil allowed the most water to seep through it? How does this compare to the predictions? Was the soil that allowed the most water to seep through, also the one that did it the fastest?

Communicate the Conclusions: On the Lab Record Card that contains the predictions, write the conclusions of this lab.

Optional: Ask someone else to predict about the 3 soil samples. Repeat the lab to check the predictions.

Spark Questions: Discuss questions sparked by this lab.

New Loop: Choose one question to investigate further, or complete a new *Investigative Loop* to measure how far water will seep in soil samples. Put the soil in a wide, clear container and measure the distance the water moves in the soil.

✎✎✎ **Design Your Own Experiment:** Select a topic based upon this *Investigative Loop* experience. See page vii for more details.

Make a Stalactite

Focus Skill: observing a concept

Activity Materials: 2 jars   water   18" of cotton string   2 paper clips
                            plate   Epsom salt

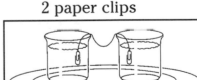

Activity: Find a location for this activity where it will not be disturbed for 2 weeks. Mix Epsom salt and water into a very saturated mixture. Pour the salt mixture into the jars. Tie paper clips on each end of the string. Place the jars on each side of a plate. Dip the string into the mixture by putting a paper clip and string into each jar. Drape the string into a slightly V-shaped design over the plate. Observe it daily.

Discuss the Activity: Describe the structure on the string. How did it grow? What does this activity tell you about stalactites?

Fascinating Facts about Earth – Graphic Organizer

Focus Skill: map reading

Paper Handouts: a copy of Graphics 17B-C
Earth Shutter Fold Project

Graphic Organizer: Cut out Graphics 17B-C and fold on the middle line so that the illustration is on the cover of each little book. Follow the directions below for the inside and glue it to the appropriate place on the world map inside the *Earth Shutter Fold Project*.

✎ Draw the cover pictures on the inside and color them.

✎✎ Copy information from the *Lots of Science Library Book* about the cover picture.

✎✎✎ Write information about the cover pictures.

Experiences, Investigations, and Research

Select one or more of the following activities for individual or group enrichment projects. Allow your students to determine the format in which they would like to report, share, or graphically present what they have discovered. This should be a creative investigation that utilizes your students' strengths.

1. Investigate the cave mineral deposits called speleothems.

2. Compare and contrast a wet cave (still in formation) and a dry cave (formation has stopped). Explain how cave systems might have both wet and dry regions.

3. Research how humans have used caves since prehistoric times.

4. Draw a diagram of a cave system. Label the sunlight zone, the twilight zone, and the midnight zone.

5. Discover and describe life forms within a cave: bats, cave crickets, mice, fish, salamanders, and more.

6. Use the Internet to investigate troglobites, or cave animals that are blind or eyeless.

7. Sketch a cave column and explain its formation.

8. The Virtual Cave
   http://www.goodearthgraphics.com/virtcave.html

Notes

How do glaciers affect the land?

Lithosphere Concepts:

- There are two kinds of glaciers: valley glaciers which form in high mountain valleys, and continental glaciers which form on ice caps in frigid polar regions.
- Layers upon layers of compacted snow form glaciers.
- Glaciers move downhill because of gravity, their weight, and melting ice underneath them.
- As glaciers move, they erode the land, often creating a U-shaped valley with steep sides and a flat floor.

Vocabulary: glacier   valley   polar regions   iceberg   *moraines   *eskers   *arête

Read: *Lots of Science Library Book #18.*

Activities:

Land Features created by a Glacier – Graphic Organizer

Focus Skills: explaining a process, labeling parts
Paper Handouts: a copy of Graphic 18A   *Landforms and Surface Features of Earth*
Graphic Organizer: Glue Graphic 18A under the previous page of *Landforms and Surface Features of Earth*.

✎ Explain what you have learned about glaciers. Color the illustrations. Draw your own examples on the left page.

✎✎ Use the *Lots of Science Library Book #18* to label the illustrations. Write clue words about glaciers: *compacted snow, moves downhill, erodes the land, creates U-shaped valley with steep sides and flat floor.*

✎✎✎ Complete ✎✎. Research the *Fascinating Physical Features of Earth* examples from the *Lots of Science Library Book #18* or other examples of glaciers. Write a descriptive or expository paragraph about them on the left page. Research and list geographic locations of several valley glaciers on the left page.

A Glacier – Investigative Loop – Lab 18-1

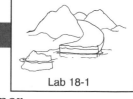

Lab 18-1

Focus Skills: demonstrating a concept, applying information
Lab Materials: dirt   plastic container for the freezer   water
Paper Handouts: a copy of Lab Graphic 18-1   8.5" x 11" sheet of paper
       Lab Record Cards   Lab Book
Graphic Organizer: Make a Pocket Book and glue it side-by-side to the Lab Book. Glue Lab Graphic 18-1 on the left pocket.
Question: How does a glacier change land features?
Research: Read *Lots of Science Library Book # 18* and review the Question.
Procedure: Put 3" of water in the plastic container. Leave it in the freezer until it is completely

frozen. Create a slightly sloped hill of dirt about 5" deep, outside or in a large container. Place the frozen water on the top of the sloped hill, pushing it slightly into the dirt.

Observations: Observe the dirt closely before the ice is placed on it. Observe the slope on a regular basis. If it takes too long to melt, pour some water on the ice.

Record the Data: Label 2 Lab Record Cards "Lab 18-1." On one card, draw the sloped hill before the ice is placed on it. On the other card, draw the sloped hill after the ice has melted. Record any features of the dirt that you observed.

Conclusions: Review the Lab Record Cards and determine how the ice changed the sloped hill. Draw conclusions about glaciers' effect on land, based on this lab.

Communicate the Conclusions: On another Lab Record Card, explain your conclusions about this lab, or write a letter to someone explaining the lab and the conclusions.

Spark Questions: Discuss questions sparked by this lab.

New Loop: Choose one question to investigate further.

✎✎✎ **Design Your Own Experiment:** Select a topic based upon this *Investigative Loop* experience. See page vii for more details.

Fascinating Facts about Earth – Graphic Organizer

Focus Skill: map reading

Paper Handouts: a copy of Graphics 18B-C   *Earth Shutter Fold Project*

Graphic Organizer: Cut out Graphics 18B-C and fold on the middle line so that the illustration is on the cover of each little book. Follow the directions below for the inside and glue it to the appropriate place on the world map inside the *Earth Shutter Fold Project.*

✎ Draw the cover pictures on the inside and color them.

✎✎ Copy information from the *Lots of Science Library Book* about the cover pictures.

✎✎✎ Write information about the cover pictures.

Experiences, Investigations, and Research

Select one or more of the following activities for individual or group enrichment projects. Allow your students to determine the format in which they would like to report, share, or graphically present what they have discovered. This should be a creative investigation that utilizes your students' strengths.

1. Investigate glacial deposits called moraines. Explain how Long Island was once a glacial moraine.

2. Draw and label a diagram of a glacier. Sketch land forms associated with glacial deposits: moraines, drumlins and eskers.

3. Explain why the majority of the world's lakes are in the northern hemisphere.

4. Investigate living organisms that have been preserved in ice. For example, the Ice Man and woolly mammoths.

5. Locate these famous glaciers on a map:
   French and Swiss Alps glaciers: Mer de Glace on Mont Blanc, Aletsch Glacier near the Jungfrau.
   Norway: Jostedal Glacier is the largest on the European continent.
   North America: Malaspina Glacier on Yakutat Bay, Alaska.

How do rivers shape the land?

Lithosphere Concepts:

- Rivers shape the land by erosion and deposition.
- The beginning of a river is called the source. A river's source is often in hills or mountains.
- River water comes from rain, melting snow, or underground springs.
- As river water flows, it carries sediment from one area to another.
- A large river has three stages: the upper stage, middle stage, and lower stage.
- Small streams or rivers that flow into a larger river are called tributaries.

Vocabulary: river source upper stage middle stage lower stage *attrition
*tributary *confluence

Read: *Lots of Science Library Book #19.*

Activities:

Rivers –Graphic Organizer

Focus Skill: describing a process
Paper Handouts: a copy of Graphic 19A *Landforms and Surface Features of Earth*
Graphic Organizer: Glue Graphic 19A under the previous page of *Landforms and Surface Features of Earth.*

✎ Explain what you have learned about rivers. Color the illustrations. Draw your own examples on the left page.

✎✎ Use the *Lots of Science Library Book #19* to label the illustrations. Write clue words about rivers: *source is the beginning, three stages, erodes rock and land, carries sediment, flows to sea or lake.*

✎✎✎ Complete ✎✎. Research the *Fascinating Physical Features of Earth* examples from the *Lots of Science Library Book #19* or other examples of rivers. Write a descriptive or expository paragraph about them on the left page.

Fascinating Facts about Earth – Graphic Organizer

Focus Skill: map reading
Paper Handouts: a copy of Graphics 19B-C *Earth Shutter Fold Project*
Graphic Organizer: Cut out Graphics 19B-C and fold on the middle line so that the illustration is on the cover of each little book. Follow the directions below for the inside and glue it to the appropriate place on the world map inside the *Earth Shutter Fold Project.*

19C

19B

✎ Draw the cover pictures on the inside and color them.

✎✎ Copy information from the *Lots of Science Library Book* about the cover pictures.

✎✎✎ Write information about the cover pictures.

Make a River

Focus Skill: demonstrating a process

Activity Materials: dirt   water

Activity: Make a sloped hill with the dirt, outside or in a plastic dish pan. Pour water in one area at the top of the hill, causing it to erode the area and begin to form banks. Continue to add water in different amounts and observe how this alters the river.

Activity Discussion: Did your created river have a source and three stages? How was it the same as a larger river? How did it differ?

Experiences, Investigations, and Research

Select one or more of the following activities for individual or group enrichment projects. Allow your students to determine the format in which they would like to report, share, or graphically present what they have discovered. This should be a creative investigation that utilizes your students' strengths.

1. Diagram and label the parts of a river.

2. Explain the importance of rivers to early civilizations: Nile River Valley, Indus River Valley, Huang-He Valley, and the Tigris-Euphrates Fertile Crescent.

3. Describe what part rivers play in each of the following: transportation, economy, recreation, and energy production.

4. How do plants and animals adapt to and survive in the moving water of a river?

5. http://wrgis.wr.usgs.gov/docs/usgsnps/noca/sb16river.html

What special features does a river form?

Lithosphere Concepts:

- As a river flows, it may travel from hard rocks to a weaker rock bed that gradually erodes away, resulting in a waterfall.
- An area may have soft rock and hard rock deposited close together. As the soft rock is eroded, the hard rock stays in place. Water flowing over these outcrops creates rapids.
- Rivers flow down slightly sloped areas, tend to wind backward and forward in a snake-like manner. As they near an ocean or a lake, they form horseshoe-shaped bends called meanders.
- When rivers flow into an ocean or a lake, they deposit the sediment they are carrying. If the sediment is deposited faster than rains or tides can wash it away, it builds up an area of flat land called a delta.

Vocabulary: waterfall   rapids   delta

Read: *Lots of Science Library Book #20.*

Activities:

Special Features of a River – Graphic Organizer

Focus Skill: describing processes

Paper Handouts: a copy of Graphics 20A   *Landforms and Surface Features of Earth*

Graphic Organizer: Glue Graphics 20A under the previous page of *Landforms and Surface Features of Earth.*

✎ Explain what you have learned about river features. Color the illustration. Draw your own examples on the left page.

✎✎ Use the *Lots of Science Library Book #20* to label the illustrations. Write clue words about river features: *hard rock, soft rock erodes, steep hill under the water.* Write clue words about deltas: *sediment is deposited, it collects, makes islands of swamps, lagoons, or sand bars.*

✎✎✎ Complete✎✎. Research the *Fascinating Physical Features of Earth* examples from the *Lots of Science Library Book #20* or other examples of waterfalls. Write a descriptive or expository paragraph about them on the left page.

Fascinating Facts about Earth – Graphic Organizer

Focus Skill: map reading

Paper Handouts: a copy of Graphics 20B-C   *Earth Shutter Fold Project*

Graphic Organizer: Cut out Graphics 20B-C and fold on the middle line so that the illustration is on the cover of each little book. Follow the directions below for the inside and glue it to the appropriate place on the world map inside the *Earth Shutter Fold Project*.

✎ Draw the cover pictures on the inside and color them.

✎✎ Copy information from the *Lots of Science Library Book* about the cover pictures.

✎✎✎ Write information about the cover pictures.

Experiences, Investigations, and Research

Select one or more of the following activities for individual or group enrichment projects. Allow your students to determine the format in which they would like to report, share, or graphically present what they have discovered. This should be a creative investigation that utilizes your students' strengths.

1. Research the Mississippi River delta. How much sediment is deposited in a day? a year? How far does the sediment reach into the Gulf of Mexico?

2. Design an activity to show how a waterfall is formed. You might use bricks, sand, and a water hose.

3. Write a fictional story about discovering an unknown waterfall.

4. Locate Iguazu Falls (EE gwah SOO) on a map, on the border between Brazil and Argentina.

5. Research the importance of the Nile River to life in Egypt both past and present.

6. http://www.infoniagara.com/d-att-american.html

How are lakes created?

Lithosphere Concepts:

- A lake is a body of water surrounded by land. Lake water comes from rivers, flooding, melting glaciers, and underground water.
- There are five main types of lakes.
  1) The deepest and most stable lakes are called rift valley lakes.
  2) A glacier lake is created when land is eroded by the weight of a glacier and then filled in as ice melts.
  3) Crater lakes are formed when volcanic craters are filled with water.
  4) Oxbow lakes are formed when the meander of a river is cut off from the rest of the river.
  5) Artificial lakes are created when people dig a hollow area in land and fill it with water.

Vocabulary: rift valley lake   glacier lake   crater lake   oxbow lake   artificial lake

Read: *Lots of Science Library Book #21.*

Activities:

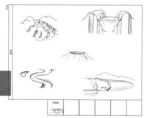

Lakes – Graphic Organizer

Focus Skill: describing a process
Paper Handouts: a copy of Graphic 21A
Graphic Organizer:

Landforms and Surface Features of Earth

- ✎ Explain what you have learned about lakes. Color the illustrations. Draw your own examples on the left page.
- ✎✎ Complete ✎. Use the *Lots of Science Library Book #21* to label the illustrations. Write clue words about lakes: *water surrounded by land, five types, rift valley, glacier, crater, oxbow, artificial.*
- ✎✎✎ Complete ✎✎. Research the *Fascinating Physical Features of Earth* examples from the *Lots of Science Library Book #21* or other examples of lakes. Write a descriptive or expository paragraph about them on the left page.

Fascinating Facts about Earth – Graphic Organizer

Focus Skill: map reading
Paper Handouts: a copy of Graphics 21B-C.   *Earth Shutter*
          Fold Project
Graphic Organizer: Cut out Graphics 21B-C and fold on the
          middle line so that the illustration is on the cover of each
          little book. Follow the directions below for the inside and

glue it to the appropriate place on the world map inside the *Earth Shutter Fold Project*.

✎ Draw the cover pictures on the inside and color them.

✎✎ Copy information from the *Lots of Science Library Book* about the cover pictures.

✎✎✎ Write information about the cover pictures.

Make a Lake

Focus Skill: demonstrating processes

Activity Materials: dirt   water

Activity: Use the dirt and water to create each type of lake. Demonstrate the process of each type of lake formation.

Activity Discussion: How was the formation of your lakes the same as that of a natural lake? How did it differ?

Experiences, Investigations, and Research

Select one or more of the following activities for individual or group enrichment projects. Allow your students to determine the format in which they would like to report, share, or graphically present what they have discovered. This should be a creative investigation that utilizes your students' strengths.

1. Locate the world's largest lakes on a map or globe. Research these lakes and make a table to record what you discover. Include each lake's location, size, temperature, and composition.

2. Research the Great Lakes. Make a time line of their history, use, and development.

3. Draw a cross-section diagram of a lake. Describe how and why deposition is taking place.

4. Compare and contrast the flora and fauna of a lake and the flora and fauna of a river.

5. Formation of Crater Lake
   http://craterlake.wr.usgs.gov/formation.html

6. The Great Lakes
   http://www.great-lakes.net/lakes/

7. What is an Oxbow Lake?
   http://mbgnet.mobot.org/fresh/lakes/oxbow.htm

How does ocean movement alter coastlines?

Lithosphere Concepts:

- Coastlines are the most rapidly changing landscapes on Earth.
- The shore area between high tide and low tide generally shows the greatest amount of erosion.
- Waves usually hit a shore at an angle and then retreat in a straight line.
- Waves break down rocks, creating formations on a rocky shore.

Vocabulary: coastline   shore   beach   cliff   tides   waves   *longshore drift
*hydraulic action

Read: *Lots of Science Library Book #22.*

Activities:

Coastlines – Graphic Organizer

Focus Skill: describing a process
Paper Handouts: a copy of Graphic 22A   *Landforms and Surface Features of Earth*
Graphic Organizer: Glue Graphic 22A under the previous page of *Landforms and Surface Features of Earth.*

✎   Explain what you have learned about coastlines. Color the illustrations. Draw your own examples on the left page.

✎✎   Use the *Lots of Science Library Book #22* to label the illustrations. Write clue words about coastlines: *tides, waves, erosion, zigzag movement on shore, arch, stack.*

✎✎✎   Complete ✎✎. Research the *Fascinating Physical Features of Earth* examples from the *Lots of Science Library Book #22* or other examples of coastlines. Write a descriptive or other expository paragraph about them on the left page.

Fascinating Facts about Earth – Graphic Organizer

Focus Skill: map reading
Paper Handouts: a copy of Graphics 22B-D
Earth Shutter Fold Project
Graphic Organizer: Cut out Graphics 22B-D and fold on the middle line so that the illustration is on the cover of each little book. Follow the directions below and glue it on the world map inside the *Earth Shutter Fold Project.*

✎   Draw the cover pictures on the inside and color them.

✎✎   Copy information from the *Lots of Science Library Book* about the cover pictures.

✎✎✎   Write information about the cover pictures.

Focus Skill: demonstrating processes

Activity Materials: dirt   water   3 rulers   thin plastic sheet or board

Activity: Either outside or in a dishpan, build a coastline with the dirt. Pour water in an area near the coastline. Use the sheet or board to move the water toward the shore in a wave-like manner.

Discuss the Activity: How is your coastline altered by the waves? How is your coastline the same as a natural one? How does it differ?

Experiences, Investigations, and Research

Select one or more of the following activities for individual or group enrichment projects. Allow your students to determine the format in which they would like to report, share, or graphically present what they have discovered. This should be a creative investigation that utilizes your students' strengths.

1. Compare and contrast the east and west coasts of the United States. Which coast is more rocky? Which coast is sandy?

2. How would United States history differ if the first explorers and colonists had arrived from the west instead of the east?

3. Research shore life. How do animals survive during high and low tides?

4. Compare natural and human-made ports.

5. List examples of countries with and without coastlines.

6. http://www.letsstudy.co.uk/student/humanities/geography/findex.html

What are the features of the ocean floor?

Lithosphere Concepts:

- The ocean floor consists of the continental shelf, the slope, and the ocean floor.
- The continental shelf and slope are part of the continents.
- The continental slope drops to the ocean floor.
- The ocean floor is a landscape of plains, mountains, volcanoes, plateaus, valleys, and trenches.

Vocabulary: continental   shelf   slope   trench   *abyssal plains

Read: *Lots of Science Library Book #23.*

Activities:

The Ocean Floor – Graphic Organizer

Focus Skill: describing a process
Paper Handouts: *Landforms and Surface Features of Earth*
Graphic Organizer: On Graphic 22A used in Lesson 22:

- ✎ Explain what you have learned about the ocean floor. Color the illustrations. Draw your own examples on the left page.
- ✎✎ Use the *Lots of Science Library Book #23* to label the illustrations. Write clue words about the ocean floor: *continental shelf, continental slope, plains, mountain chain, ridges, trenches.*
- ✎✎✎ Complete ✎✎. Research the *Fascinating Physical Features of Earth* examples from the *Lots of Science Library Book #23* or other examples of features of the ocean floor. Write a descriptive or expository paragraph about them on the left page.

Fascinating Facts about Earth – Graphic Organizer

Focus Skill: map reading
Paper Handouts: a copy of Graphics 23B-C   *Earth Shutter Fold Project*
Graphic Organizer: Cut out Graphics 23B-C and fold on the middle line so that the illustration is on the cover of each little book. Follow the directions below for the inside and glue it to the appropriate place on the world map inside the *Earth Shutter Fold Project*.

- ✎ Draw the cover pictures on the inside and color them.
- ✎✎ Copy information from the *Lots of Science Library Book* about the cover pictures.
- ✎✎✎ Write information about cover pictures.

23B

23C

Experiences, Investigations, and Research

Select one or more of the following activities for individual or group enrichment projects. Allow your students to determine the format in which they would like to report, share, or graphically present what they have discovered. This should be a creative investigation that utilizes your students' strengths.

1. Diagram and label a cross-section of the ocean. Show the shelf, slope, and floor.

2. Compare and contrast the shelf and slope off the east and west coasts of the United States.

3. Research the levels of life in the ocean. Are there more life forms over the shelf and slope, or over the ocean floor? Why?

4. Sketch the Mariana Trench and draw Mt. Everest submerged in the trench. How much water would be over the top of Mt. Everest?

5. Make a time line outlining the history of ocean floor exploration.

6. Explain how and why the floor of the Atlantic Ocean is expanding and the floor of the Pacific Ocean is shrinking by a few centimeters each year.

7. http://www.onr.navy.mil/focus/ocean/regions/oceanfloor1.htm

What do we know about deserts?

Lithosphere Concepts:

- Deserts are dry, barren areas with low rainfall, high evaporation, and drastic temperature changes from day to night.
- One-quarter of Earth's land is desert.
- In a dry environment like a desert, wind is the main factor in shaping the landscape.
- In sandy deserts, wind shapes and moves the sand, forming dunes.
- Because the continent of Antarctica is covered in ice and does not receive much precipitation each year, it is considered a desert.

Teacher's Note: An alternative assessment suggestion for this lesson is found on pages 64-65. If Graphics Pages are being consumed, first photocopy assessment graphics that are needed.

Vocabulary: desert   barren   wind   *zeugen   *barchans   *seif dunes

Read: *Lots of Science Library Book #24.*

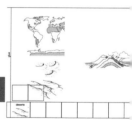

Activities:

Deserts – Graphic Organizer

Focus Skill: describing a process
Paper Handouts: a copy of Graphics 24A   *Landforms and Surface Features of Earth*
Graphic Organizer: Glue Graphic 24A under the previous page in *Landforms and Surface Features of Earth.*

✎ Explain what you have learned about deserts. Color the illustrations. Draw your own examples on the left page.

✎✎ Use the *Lots of Science Library Book #24* to label the illustrations. Write clue words about deserts: *dry, barren, sandy desert is altered by the wind; icy desert is interior of Antarctica.*

✎✎✎ Complete ✎✎. Research the *Fascinating Physical Features of Earth* examples from the *Lots of Science Library Book #24* or other examples of deserts. Write a descriptive or expository paragraph about them on the left page.

Fascinating Facts about Earth – Graphic Organizer

Focus Skill: map reading
Paper Handouts: a copy of Graphics 24B-C
Earth Shutter Fold Project
Graphic Organizer: Cut out Graphics 24B-C and fold on the
middle line so that the illustration is on the cover of each

24B

24C

little book. Follow the directions below for the inside and glue it to the appropriate place on the world map inside the *Earth Shutter Fold Project.*

✎ Draw the cover pictures on the inside and color them.

✎✎ Copy information from the *Lots of Science Library Book* about the cover pictures.

✎✎✎ Write information about the cover pictures.

Experiences, Investigations, and Research

Select one or more of the following activities for individual or group enrichment projects. Allow your students to determine the format in which they would like to report, share, or graphically present what they have discovered. This should be a creative investigation that utilizes your students' strengths.

1. Compare and contrast the hot, interior desert of Australia and the frozen interior desert of Antarctica.

2. Locate the Andes Mountains on a map or globe and describe their rainshadow. Are there any deserts to the east of the mountains?

3. Locate Earth's deserts on a map or globe. Make a table of information on several deserts. Include size, location, average day and night temperatures, and average yearly precipitation amounts.

4. Research the flora and fauna of a desert of your choice.

5. Sketch the different types of dunes and describe their formation.

6. Locate the Sahara Desert on a map or globe. Research its size; past and present.

7. Describe how deserts can be transformed into crop land.

8. http://pubs.usgs.gov/gip/deserts/contents/

Assessment: An Ongoing Process

Students do not have to memorize every vocabulary word or fact presented in these science lessons. It is more important to teach them general science processes and cause and effect relationships. Factual content is needed for students to understand processes, but it should become familiar to them through repeated exposure, discussion, reading, research, presentations, and a small amount of memorization. You can determine the amount of content your students have retained by asking specific questions that begin with the following words: *name, list, define, label, identify, draw,* and *outline.*

Try to determine how much content your students have retained through discussions. Determine how many general ideas, concepts, and processes your students understand by asking them to describe or explain them. Ask leading questions that require answers based upon thought and analysis, not just facts. Use the following words and phrases as you discuss and evaluate: *why, how, describe, explain, determine,* and *predict.* Questions may sound like this:

What would happen if _____? *Compare _____ to _____.*
Why do you think _____ happens? *What does ___have in common with __?*
What do you think about _____? *What is the importance of _____?*

Alternative Assessment Strategies

If you need to know specifically what your students have retained or need to assign your students a grade for the content learned in this program, we suggest using one of the following assessment strategies.

By the time your students have completed a lesson in this program, they will have written about, discussed, observed, and discovered the concepts of the lesson. However, it is still important for you to review the concepts that you are assessing prior to the assessment. By making your students aware of what you expect them to know, you provide a structure for their preparations for the assessments.

1) At the end of each lesson, ask your students to restate the concepts taught in the lesson. For example, if they have made a 4 Door Book showing the steps of pollination, ask them to tell you about each step using the pictures as a prompt. This assessment can be done by you or by a student.

2) At the end of each lesson, ask your students to answer the questions on the inside back cover of the *Lots of Science Library Book* for that lesson. The answers to these questions may be done verbally or in writing. Ask older students to use their vocabulary words in context as they answer the questions. This is a far more effective method to determine their knowledge of the vocabulary words than a matching or multiple choice test on the words.

3) Provide your students with Landform Concept Maps partially completed. Ask them to fill in the blanks. Example for Lesson 8:

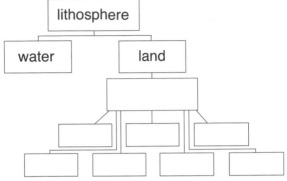

4) Use the 3D Graphic Organizers to assess your students' understanding of the concepts. Use the *Lithosphere Concepts* listed on the teachers' pages to determine exactly what you want covered in the assessments. Primary and beginning students may use the pictures found on the Graphics Pages as guides for their assessments. By using the pictures, your students are sequencing and matching, while recalling information. Older students should draw their own pictures and use their vocabulary words in their descriptions of the concepts. Below are suggestions for this method of assessment.

a) Lesson 3 – Make a Layered Look Book. Write *Earth* on the top section. On the first tab write *Earth in Space*, on the next tab write *Features of Earth*, and on the bottom tab write *Lithosphere*. Open each tab and write information about the topic.

b) Lesson 8 – Make two Small Question and Answer Books. Label each tab accordingly: *faults, folds, earthquakes, mountains, volcanoes, tsunamis, hot springs and geysers, volcanic islands*. Under each tab list information or make a drawing of each topic.

c) Lesson 14 – Make a Large Question and Answer Book. On the left tab write *Rock Cycle* and under the tab draw and label the rock cycle. On the right tab, choose one of the following to label and write about: *minerals, fossils, soil, or mineral resources*.

d) Lesson 16 – Make a Large Question and Answer Book. Label the left tab *weathering* and the right tab *erosion*. Under each tab explain the process and examples of their effect on Earth.

e) Lesson 24 – Make two Small Question and Answer Books. Label each tab accordingly: *caves, glaciers, rivers, river features, lakes, coastlines, ocean floor, deserts*. Under each tab list information or make a drawing of each topic.

Notes

Lots of Science Library Books

Each *Lots of Science Library Book* is made up of 16 inside pages, plus a front and back cover. All the covers to the *Lots of Science Library Books* are located at the front of this section. The covers are followed by the inside pages of the books.

How to Photocopy the *Lots of Science Library Books*

Note: These pages are easier to photocopy if they are taken out of the book. The *Lots of Science Library Books* are provided as consumable pages which may be cut out of the *Great Science Adventures* book at the line on the top of each page. If, however, you wish to make photocopies for your students, you can do so by following the instructions below.

Be sure to try one book before you copy the entire set. To photocopy the inside pages of the *Lots of Science Library Books*:

1. Note that there is a "Star" above the line at the top of each *LSLB* sheet.

2. Locate the *LSLB* sheet that has a Star on it above page 16. Position this sheet on the glass of your photocopier so the side of the sheet which contains page 16 is facing down, and the Star above page 16 is in the left corner closest to you. Photocopy the page.

3. Turn the *LSLB* sheet over so that the side of the *LSLB* sheet containing page 6 is now face down. Position the sheet so the Star above page 6 is again in the left corner closest to you.

4. Insert the previously photocopied paper into the copier again, inserting it face down, with the Star at the end of the sheet that enters the copier last. Photocopy the page.

5. Repeat steps 1 through 4, above, for each *LSLB* sheet.

To photocopy the covers of the *Lots of Science Library Books*:

1. Insert "Cover Sheet A" in the photocopier with a Star positioned in the left corner closest to you, facing down. Photocopy the page.

2. Turn "Cover Sheet A" over so that the side you just photocopied is now facing you. Position the sheet so the Star is again in the left corner closest to you, facing down.

3. Insert the previously photocopied paper into the copier again, inserting it face down, with the Star entering the copier last. Photocopy the page.

4. Repeat steps 1 through 3, above, for "Cover Sheets" B, C, D, E, and F.

Note: The owner of this book has permission to photocopy the *Lots of Science Library Book* pages and covers for classroom use only.

How to assemble the *Lots of Science Library Books*

Once you have made the photocopies or cut the consumable pages out of this book, you are ready to assemble your *Lots of Science Library Books*. To do so, follow these instructions:

1. Cut each sheet, both covers and inside pages, on the solid lines.

2. Lay the inside pages on top of one another in this order: pages 2 and 15, pages 4 and 13, pages 6 and 11, pages 8 and 9.

3. Fold the stacked pages on the dotted line, with pages 8 and 9 facing each other.

4. Turn the pages over so that pages 1 and 16 are on top.

5. Place the appropriate cover pages on top of the inside pages, with the front cover facing up.

6. Staple on the dotted line in two places.

You now have completed *Lots of Science Library Books*.

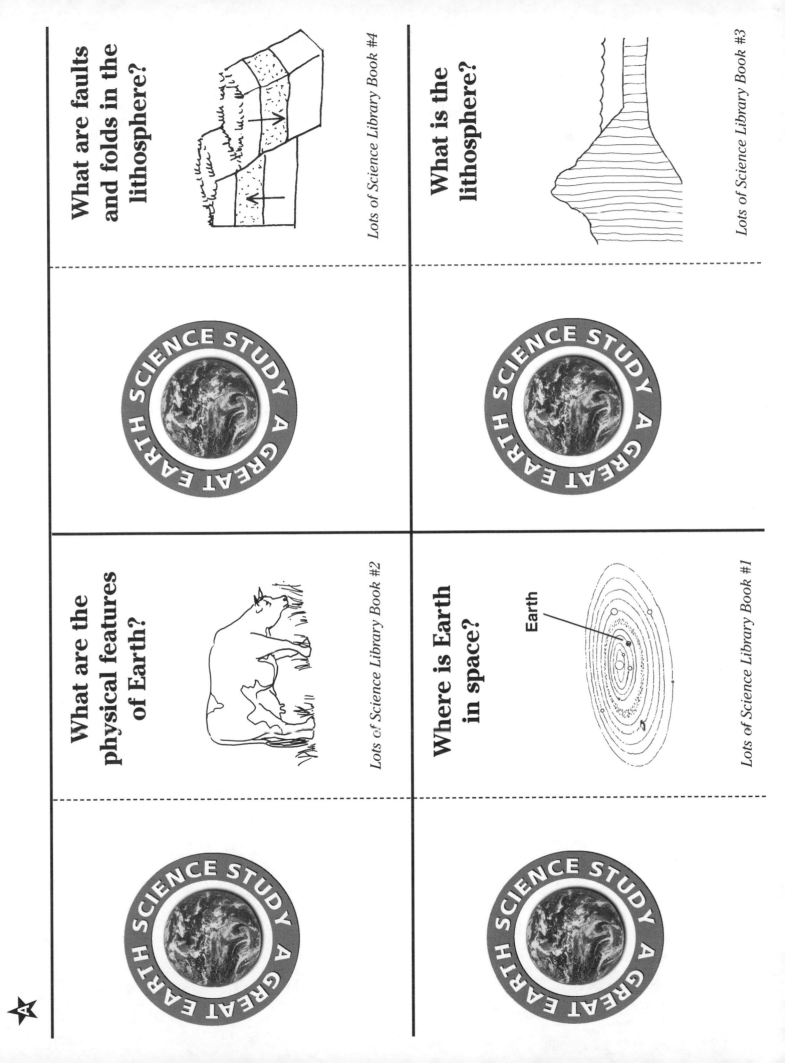

What are faults and folds in the lithosphere?

Lots of Science Library Book #4

What is the lithosphere?

Lots of Science Library Book #3

SCIENCE STUDY A GREAT EARTH

SCIENCE STUDY A GREAT EARTH

What are the physical features of Earth?

Lots of Science Library Book #2

Where is Earth in space?

Earth

Lots of Science Library Book #1

SCIENCE STUDY A GREAT EARTH

SCIENCE STUDY A GREAT EARTH

fault
fold
*anticline
*syncline

ocean
continent
*atmosphere

lithosphere
continental plates
*plate tectonics

Earth
Sun
planet
*revolution
*rotation

Explain the types of faults
and how they occur.

Explain the types of folds
and how they occur.

Describe the surface of
Earth.

Explain the importance of
Earth's atmosphere.

Describe Earth's lithosphere.

Explain how continental
plates slide.

Describe Earth in space.

What are tsunamis, hot springs, geysers, and volcanic islands?

Lots of Science Library Book #8

What is a volcano?

Lots of Science Library Book #7

SCIENCE STUDY A GREAT EARTH

SCIENCE STUDY A GREAT EARTH

How are mountains formed?

Lots of Science Library Book #6

What are earthquakes?

Lots of Science Library Book #5

SCIENCE STUDY A GREAT EARTH

SCIENCE STUDY A GREAT EARTH

B

hot springs
geyser
volcanic
*tsunamis

volcano
magma
lava
vent
*eruption
*dormant
*extinct

Explain how a tsunami is formed.

Explain how hot springs and geysers are formed.

Describe the four classes of volcanoes.

Explain how a volcano is formed.

mountain
folded
block
dome
volcanic

Describe the four different types of mountains and how they are formed.

earthquake
focus
vibrate
aftershocks
*Richter scale
*Mercalli scale

Describe the scales that are used to measure earthquakes.

Explain how earthquakes occur.

What are fossils?

SCIENCE STUDY A GREAT EARTH

What are minerals?

SCIENCE STUDY A GREAT EARTH

What are metamorphic rocks?

SCIENCE STUDY A GREAT EARTH

What are igneous and sedimentary rocks?

SCIENCE STUDY A GREAT EARTH

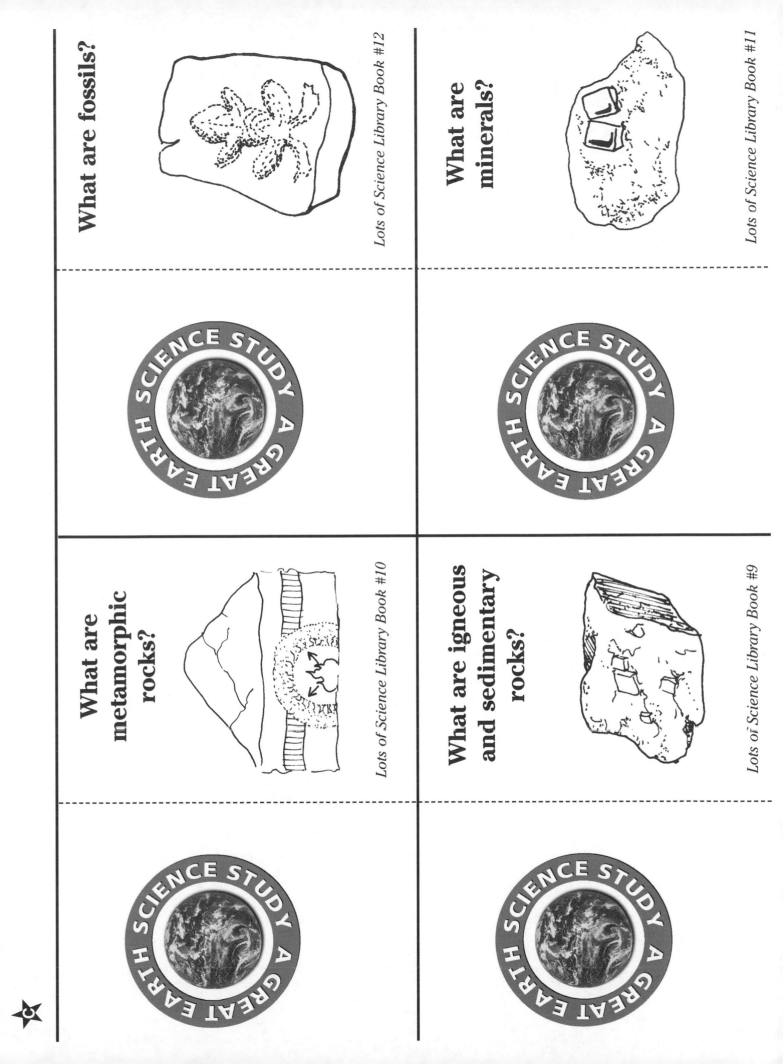

fossil
decay

*fossilization
*petrified

minerals
elements

*Mohs scale
*native element
*compound element

Explain how fossils are
formed.

Describe minerals.

Explain the two groups of
minerals.

metamorphic
rock cycle

igneous
sedimentary

*organic
*inorganic

Explain how metamorphic
rocks are formed.

Explain the rock cycle.

Explain how igneous rocks
are formed.

Explain how sedimentary
rocks are formed.

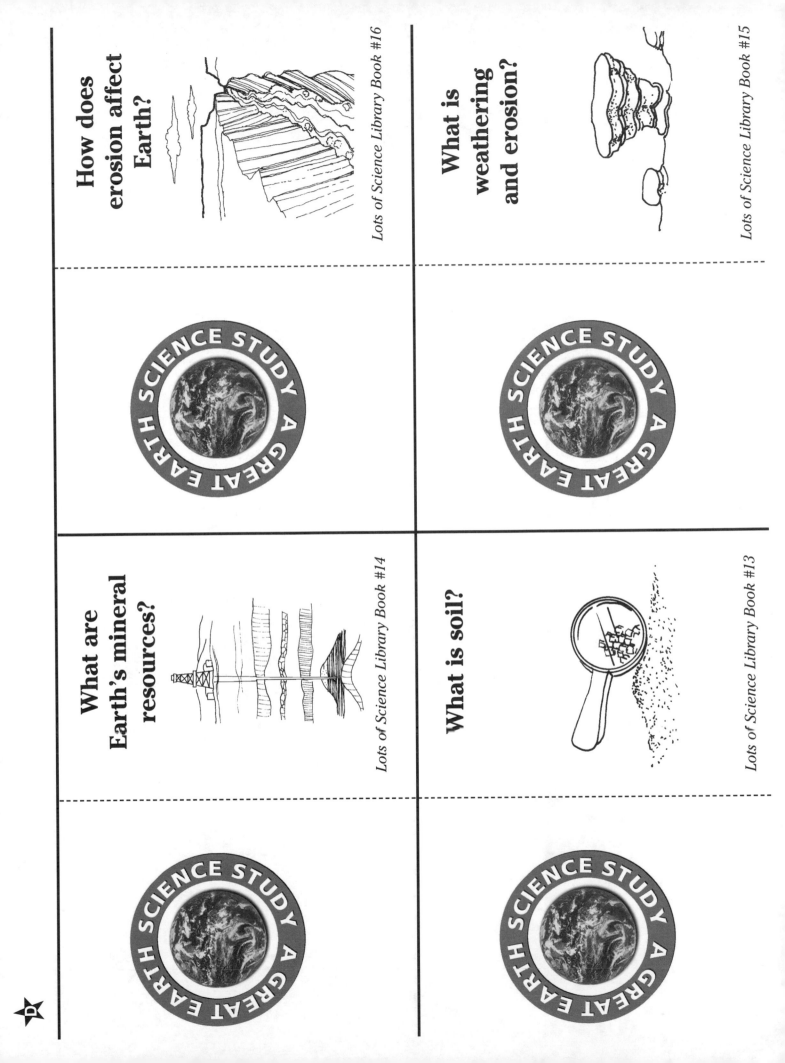

How does erosion affect Earth?

Lots of Science Library Book #16

What is weathering and erosion?

Lots of Science Library Book #15

What are Earth's mineral resources?

Lots of Science Library Book #14

What is soil?

Lots of Science Library Book #13

SCIENCE STUDY A GREAT EARTH

D ★

canyon
mesa
butte
*monadnock

List four land features that are created by erosion.

resources
fuel
coal
oil
gas
*chemically
*peat
*lignite

Describe several of Earth's natural resources.

Explain how fuels and raw materials are formed in Earth's lithosphere.

weathering
physical
chemical
erosion
debris

Explain three ways that rocks are weathered.

soil
climate
geography

Describe the five main factors that affect soil.

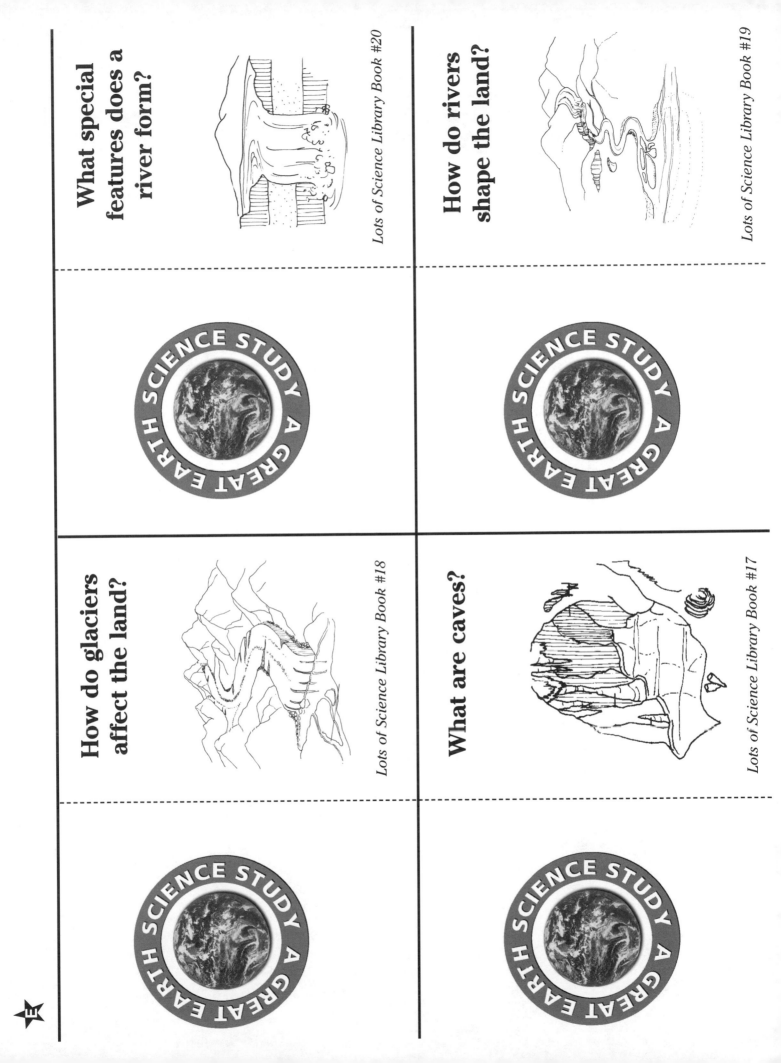

What special features does a river form?

Lots of Science Library Book #20

How do rivers shape the land?

Lots of Science Library Book #19

SCIENCE STUDY A GREAT EARTH

SCIENCE STUDY A GREAT EARTH

How do glaciers affect the land?

Lots of Science Library Book #18

What are caves?

Lots of Science Library Book #17

SCIENCE STUDY A GREAT EARTH

SCIENCE STUDY A GREAT EARTH

waterfall
rapids
delta

Explain how each of these features are formed: waterfall, rapids, and delta.

river
source
upper stage
middle stage
lower stage
*attrition
*tributary
*confluence

Explain how rivers are formed.

Describe the three stages of a river and how each shapes the land.

glacier
valley
polar regions
iceberg
*moraines
*eskers
*arête

Explain how glaciers create land formations.

cave
limestone
structures
evaporate
pillar
*stalactite
*stalagmite

Explain how caves are formed underground.

What do we know about deserts?

Lots of Science Library Book #24

What are features of the ocean floor?

Lots of Science Library Book #23

How does ocean movement alter coastlines?

Lots of Science Library Book #22

How are lakes created?

Lots of Science Library Book #21

SCIENCE STUDY A GREAT EARTH

SCIENCE STUDY A GREAT EARTH

SCIENCE STUDY A GREAT EARTH

SCIENCE STUDY A GREAT EARTH

desert
barren
wind
*zeugen
*barchans
*seif dunes

continental
shelf
slope
trench
*abyssal plains

Describe a desert.
List the three largest deserts.

Describe the continental land and ocean floor.

coastline
shore
beach
cliff
tides
waves
*longshore drift
*hydraulic action

rift valley lake
glacier lake
crater lake
oxbow lake
artificial lake

Explain why coastlines are the most rapidly changing landscapes on Earth.

List the five types of lakes.
Explain how each type of lake is formed.

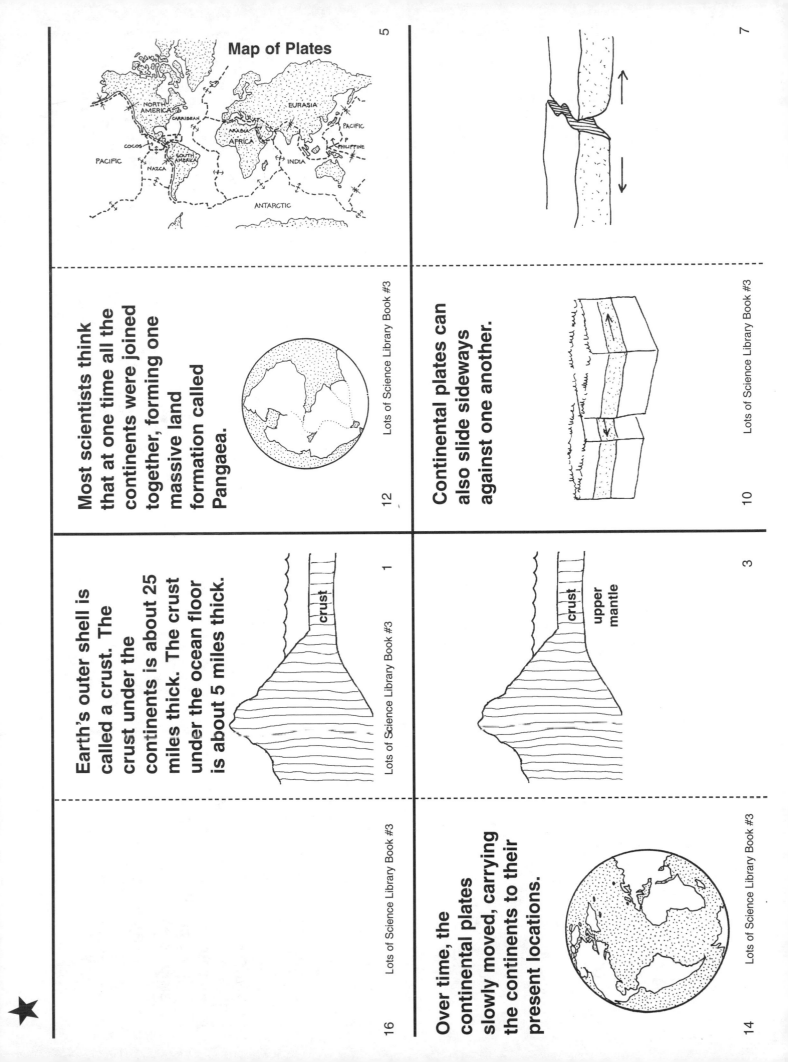

Map of Plates

5

Most scientists think that at one time all the continents were joined together, forming one massive land formation called Pangaea.

12

Lots of Science Library Book #3

Continental plates can also slide sideways against one another.

10

7

Lots of Science Library Book #3

Earth's outer shell is called a crust. The crust under the continents is about 25 miles thick. The crust under the ocean floor is about 5 miles thick.

crust

1

Lots of Science Library Book #3

16

crust

upper mantle

3

Over time, the continental plates slowly moved, carrying the continents to their present locations.

14

Lots of Science Library Book #3

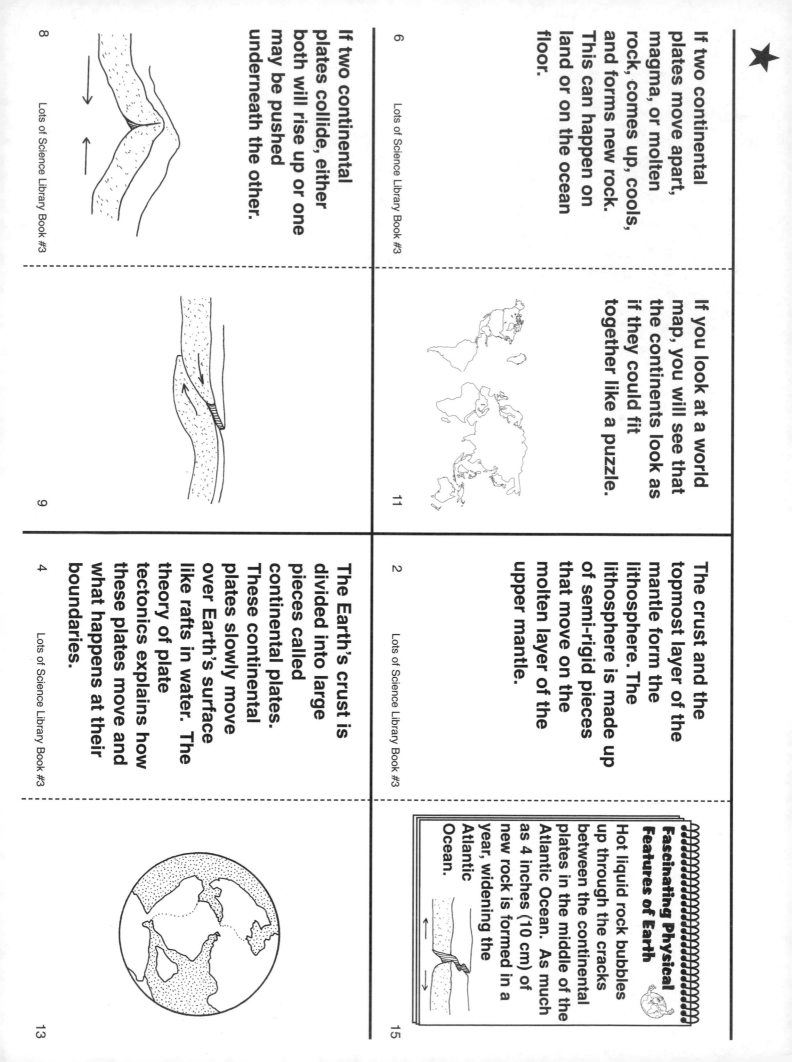

If two continental plates move apart, magma, or molten rock, comes up, cools, and forms new rock. This can happen on land or on the ocean floor.

Lots of Science Library Book #3

If you look at a world map, you will see that the continents look as if they could fit together like a puzzle.

The crust and the topmost layer of the mantle form the lithosphere. The lithosphere is made up of semi-rigid pieces that move on the molten layer of the upper mantle.

Lots of Science Library Book #3

Fascinating Physical Features of Earth

Hot liquid rock bubbles up through the cracks between the continental plates in the middle of the Atlantic Ocean. As much as 4 inches (10 cm) of new rock is formed in a year, widening the Atlantic Ocean.

Lots of Science Library Book #3

If two continental plates collide, either both will rise up or one may be pushed underneath the other.

The Earth's crust is divided into large pieces called continental plates. These continental plates slowly move over Earth's surface like rafts in water. The theory of plate tectonics explains how these plates move and what happens at their boundaries.

Lots of Science Library Book #3

A normal fault is created when pressure builds up and one side of the fault line moves upward.

5

Tear faults occur when pressure forces the pieces of crust to tear apart, or to slide sideways in opposite directions.

7

The second is a downfold.

Downfolds are called synclines.

12   Lots of Science Library Book #4

Sometimes rocks do not break when pressure builds up underneath them. These rocks will bend or crumple instead.

10   Lots of Science Library Book #4

As Earth's continental plates move together, apart, or sideways, they generate great amounts of pressure that may bend or break the rock layers forming them.

1

Lots of Science Library Book #4

The Himalayan mountain range is a giant fold in Earth's lithosphere.

3

Lots of Science Library Book #4

Fascinating Physical Features of Earth

The San Andreas Fault in California is a large crack 700 miles (1,126 km) long.

16   Lots of Science Library Book #4

These movements in Earth's crust create mountains, valleys, and other physical features on the surface. The Grand Teton Mountains were formed from a normal fault.

A reverse fault is created when the crust is squeezed, and one side of the fault line moves down.

Two types of folds may occur in Earth's lithosphere. The first is an upfold.

Upfolds are called anticlines.

Grand Tetons

Faults are cracks in the lithosphere. Major faults are located at the continental plate boundaries. Fault lines are a point of weakness in Earth's surface.

The pressure of moving continental plates can break hard, rigid rock layers, resulting in a fault.

The Himalayas

Folds vary in size from a few inches to folded mountain ranges over 100 miles long.

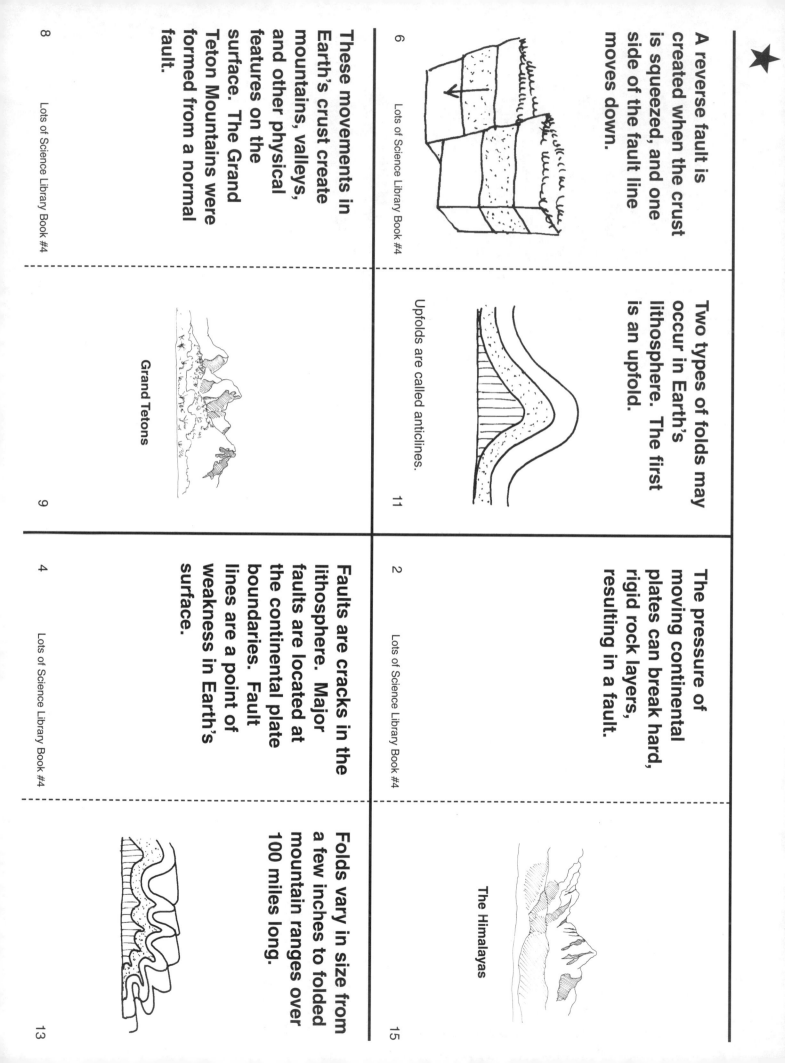

Earthquakes are most likely to occur at continental plate boundaries and fault lines.

Fascinating Physical Features of Earth

The San Francisco earthquake of 1989 measured 7.1 on the Richter scale. It destroyed homes and businesses as well as sections of the Bay Bridge and the interstate highway system.

The Richter scale uses a scale of 1 to 10, with 1 being the least powerful and 10 the most powerful, to measure earthquakes. Each step on the scale is 20 times greater than the one before it.

Any vibration in Earth's surface is considered an earthquake. We do not feel most earthquakes because they are tiny; however, when a continental plate slips suddenly, serious earthquakes can occur.

epicenter

focus

Fascinating Physical Features of Earth

An earthquake usually lasts for less than one minute. In 1755, an earthquake in Lisbon, Portugal lasted for about 10 minutes. The shock waves of this earthquake were felt as far away as North Africa.

The Mercalli Scale
1. Not felt.
2. Felt by a few people.
3. Hanging objects sway.
4. Windows rattle.
5. Liquids spill, objects fall.
6. Objects fall from walls.

Scientists who study earthquakes are called seismologists.

Scientists who study earthquakes use two different scales to measure them. The Richter scale is based on the amount of energy produced at the focus of an earthquake. It is measured with a seismometer, or a pendulum-type instrument that measures surface vibrations.

When the continental plates stop moving, rocks resettle along the fault line, causing aftershocks that vibrate throughout the area.

★

Earth experiences over 800,000 earthquakes each year that are unnoticeable and measure 2.0 – 3.4 on the Richter scale. Another 30,000 or more earthquakes are barely felt, measuring 3.5 – 4.2 on the Richter scale.

Vibrations called seismic waves travel upward and outward from the focus, causing Earth's surface to move and sometimes to crack open. In 1964, an earthquake in Alaska caused huge cracks to open in the ground, up to 3 feet wide and 40 feet deep (90 cm wide and 12 m deep).

The point in the lithosphere where the slip occurs is called the focus of the earthquake. The epicenter is the point on Earth's surface directly above the focus.

7. Buildings damaged.
8. Chimneys collapse.
9. Cracks appear in ground.
10. Severe damage to buildings.
11. Railway lines bend/underground pipes break.
12. Nearly everything is damaged.

The second scale for measuring earthquakes is called the Mercalli scale. This scale is based on eyewitness accounts and ranges from 1–12. The San Francisco earthquake of 1989 measured 10.7 on the Mercalli scale.

The Alps are an example of a folded mountain range.

The Alps

5

The Grand Teton Mountains in Wyoming are examples of block mountains.

Grand Tetons

7

Over the years, mountains change. Frost, ice, and water split and wear away the rock of a mountain. Some scientists think that mountains lose about 3.5 inches (8.6 cm) of their surface every 1,000 years.

12

10

A mountain is any point on Earth's surface that rises 1,000 feet or more above its surroundings.

1

Mountains can be steep, or they can gently slope to their height. They can be barren or covered with trees. Mountains exist on the land and under the oceans. There are several general types of mountains - folded, block, dome, and volcanic.

3

16

Hills are high points on Earth's surface that are less than 1,000 feet in height. Most hills are formed by the same processes that form mountains, while a few are formed by glacier movement and deposits.

You will learn about glaciers in *Lots of Science Library Book #18.*

14

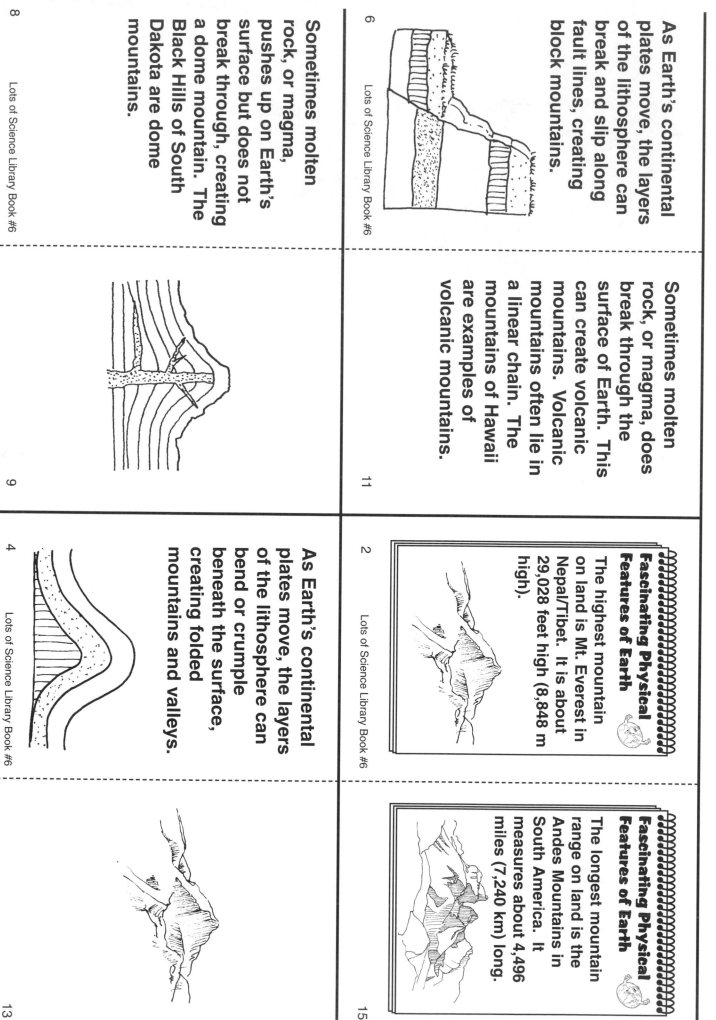

Sometimes molten rock, or magma, pushes up on Earth's surface but does not break through, creating a dome mountain. The Black Hills of South Dakota are dome mountains.

★

As Earth's continental plates move, the layers of the lithosphere can break and slip along fault lines, creating block mountains.

Sometimes molten rock, or magma, does break through the surface of Earth. This can create volcanic mountains. Volcanic mountains often lie in a linear chain. The mountains of Hawaii are examples of volcanic mountains.

As Earth's continental plates move, the layers of the lithosphere can bend or crumple beneath the surface, creating folded mountains and valleys.

2

Fascinating Physical Features of Earth

The highest mountain on land is Mt. Everest in Nepal/Tibet. It is about 29,028 feet high (8,848 m high).

Fascinating Physical Features of Earth

The longest mountain range on land is the Andes Mountains in South America. It measures about 4,496 miles (7,240 km) long.

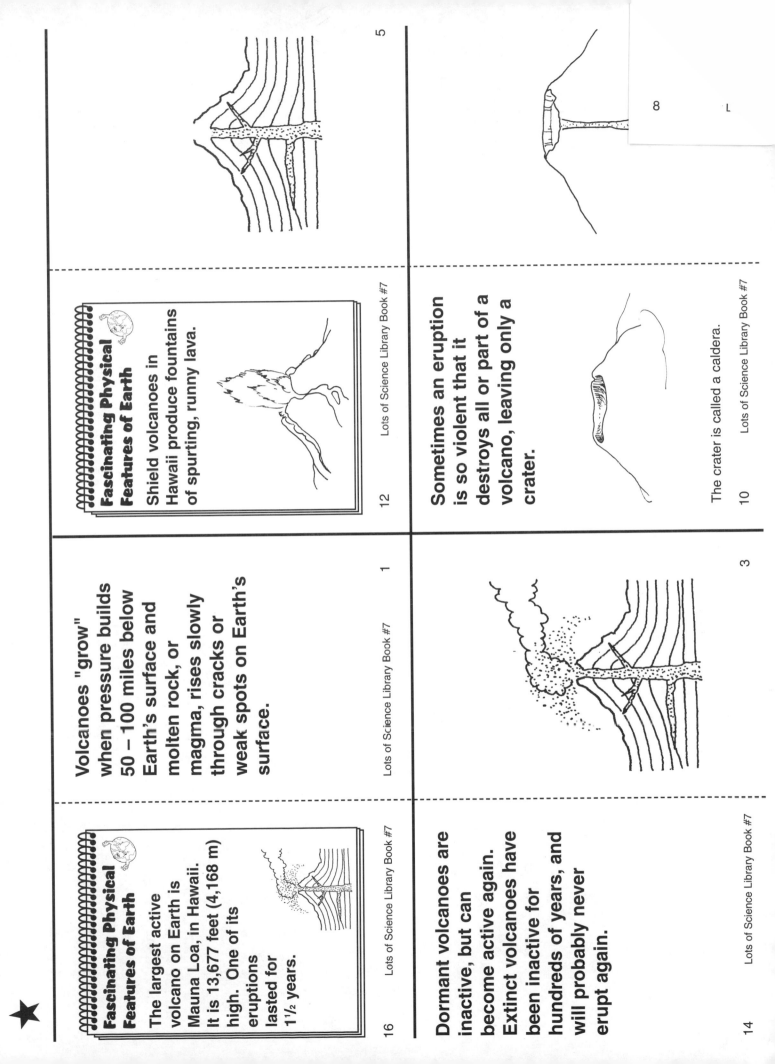

5

Fascinating Physical Features of Earth

Shield volcanoes in Hawaii produce fountains of spurting, runny lava.

12 Lots of Science Library Book #7

8   L

Sometimes an eruption is so violent that it destroys all or part of a volcano, leaving only a crater.

The crater is called a caldera.

10 Lots of Science Library Book #7

Volcanoes "grow" when pressure builds 50 – 100 miles below Earth's surface and molten rock, or magma, rises slowly through cracks or weak spots on Earth's surface.

1 Lots of Science Library Book #7

Fascinating Physical Features of Earth

The largest active volcano on Earth is Mauna Loa, in Hawaii. It is 13,677 feet (4,168 m) high. One of its eruptions lasted for 1½ years.

16 Lots of Science Library Book #7

3 Lots of Science Library Book #7

Dormant volcanoes are inactive, but can become active again. Extinct volcanoes have been inactive for hundreds of years, and will probably never erupt again.

14 Lots of Science Library Book #7

★

Non-viscous lava is thin, runny lava that does not cool quickly. This lava can travel several miles (kilometers) before it cools. This lava forms low, flat mountains called shield volcanoes.

Other eruptions are quiet and not violent. The resulting lava might be runny and flow gently from the vent.

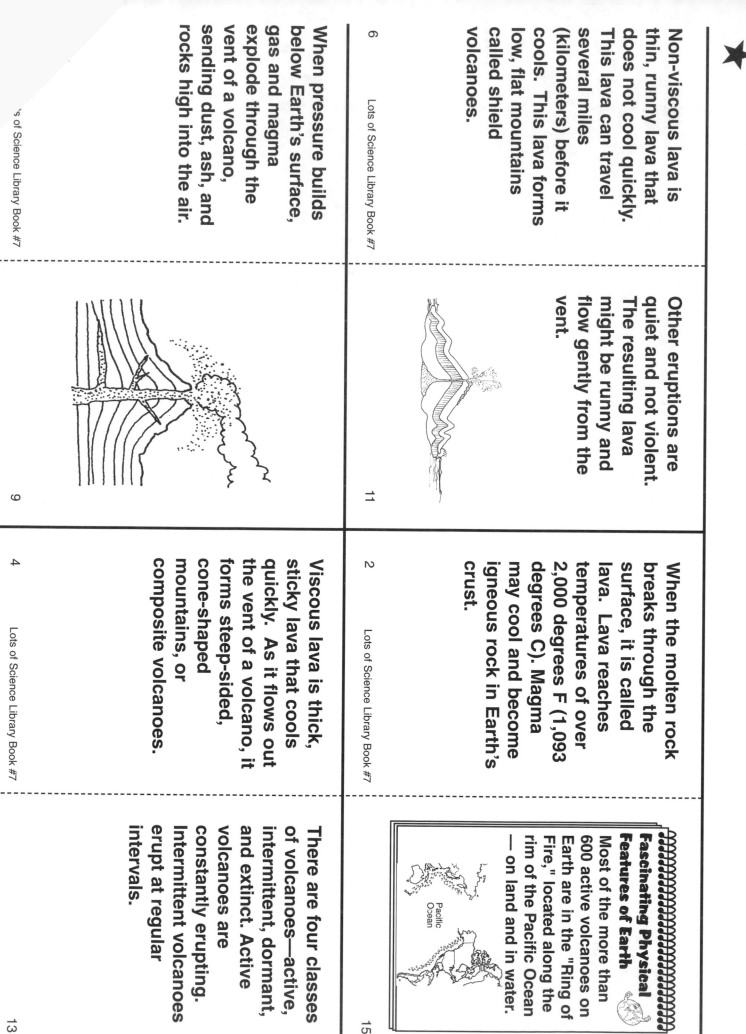

When pressure builds below Earth's surface, gas and magma explode through the vent of a volcano, sending dust, ash, and rocks high into the air.

When the molten rock breaks through the surface, it is called lava. Lava reaches temperatures of over 2,000 degrees F (1,093 degrees C). Magma may cool and become igneous rock in Earth's crust.

Viscous lava is thick, sticky lava that cools quickly. As it flows out the vent of a volcano, it forms steep-sided, cone-shaped mountains, or composite volcanoes.

There are four classes of volcanoes—active, intermittent, dormant, and extinct. Active volcanoes are constantly erupting. Intermittent volcanoes erupt at regular intervals.

Fascinating Physical Features of Earth

Most of the more than 600 active volcanoes on Earth are in the "Ring of Fire," located along the rim of the Pacific Ocean — on land and in water.

Pacific Ocean

Fascinating Physical Features of Earth

The highest tsunami recorded hit the Ishigaki Island of Japan at almost 280 feet (85 m) high.

When the water is extremely hot, it bubbles out of cracks in the crust, forming hot springs.

Volcanoes erupt under the ocean as well as on land. Sometimes a volcanic mountain "grows" high enough to be visible near the ocean surface as a volcanic island.

Geysers emit streams of hot water. As some of this water evaporates, minerals are deposited around the vent of the geyser.

Volcanic eruptions and earthquakes often occur in the ocean crust. The shock waves of a large disturbance can cause giant walls of water, called tsunamis, to form.

Fascinating Physical Features of Earth

The newest volcanic island, spotted in 1979, is Lateiki, off the east coast of Australia.

Volcanic activity in Earth's continental crust produces hot zones of mantle near Earth's surface. In some of these areas, hot rocks and magma heat underground water.

In some places, the underground water is trapped. As the water heats, pressure builds. Finally, the pressure becomes so strong that water bursts through the surface, forming a geyser. Geysers stop when the pressure is released, but they may build up again and again.

9

Fascinating Physical Features of Earth

Old Faithful, a geyser in Yellowstone National Park, erupts to a height of 130 feet (40 m) every 30-90 minutes.

11

Tsunamis are sometimes incorrectly called tidal waves.

Tsunamis are not very noticeable at sea, but when they hit land the wave can be 6 – 60 feet (1.8 – 18 m) high. Since they can travel at speeds of 500 – 600 mph (805-965 kmph), their impact can be devastating.

15

Fascinating Physical Features of Earth

One tsunami took over 4 1/2 hours to travel 2,000 miles (3,220 km) from the north Pacific to Honolulu, Hawaii. The tsunami hit the island with waves over 50 feet (15 m) high.

Not all islands are formed in this manner, but the Hawaiian Islands are an example of volcanic islands.

13

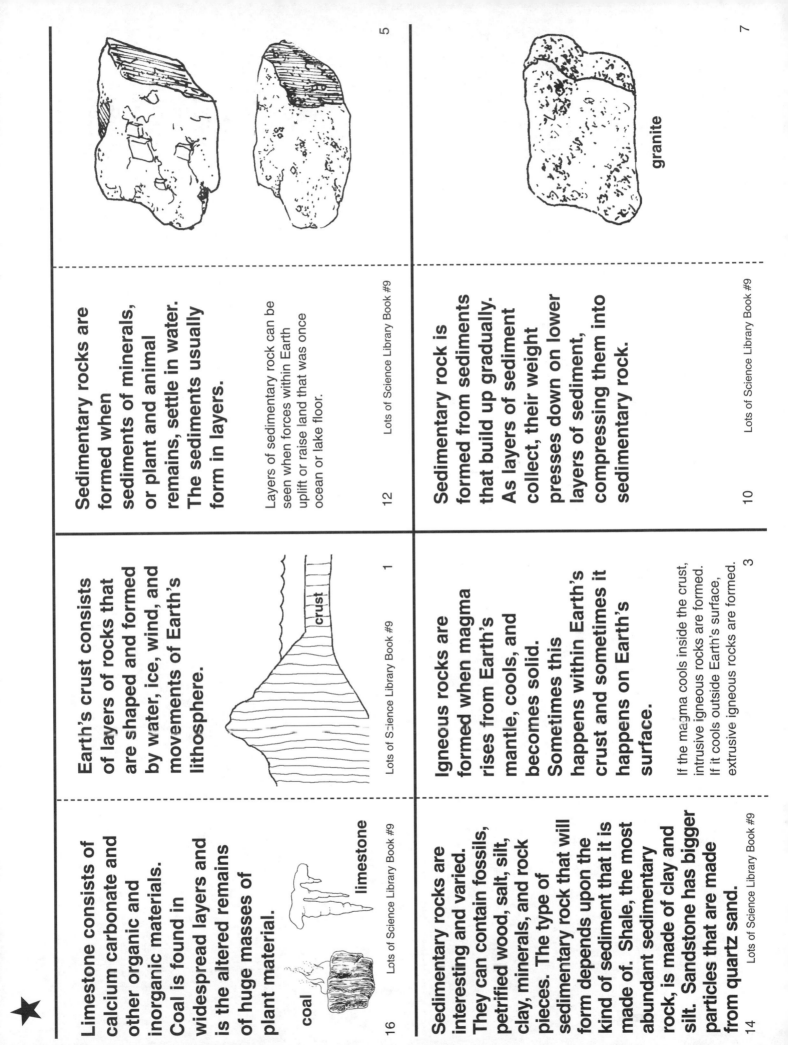

Page 5

granite — **Page 7**

Sedimentary rocks are formed when sediments of minerals, or plant and animal remains, settle in water. The sediments usually form in layers.

Layers of sedimentary rock can be seen when forces within Earth uplift or raise land that was once ocean or lake floor.

12   Lots of Science Library Book #9

Sedimentary rock is formed from sediments that build up gradually. As layers of sediment collect, their weight presses down on lower layers of sediment, compressing them into sedimentary rock.

10   Lots of Science Library Book #9

Earth's crust consists of layers of rocks that are shaped and formed by water, ice, wind, and movements of Earth's lithosphere.

crust

Lots of Science Library Book #9   1

Igneous rocks are formed when magma rises from Earth's mantle, cools, and becomes solid. Sometimes this happens within Earth's crust and sometimes it happens on Earth's surface.

If the magma cools inside the crust, intrusive igneous rocks are formed. If it cools outside Earth's surface, extrusive igneous rocks are formed.

3

Limestone consists of calcium carbonate and other organic and inorganic materials. Coal is found in widespread layers and is the altered remains of huge masses of plant material.

coal   limestone

16   Lots of Science Library Book #9

Sedimentary rocks are interesting and varied. They can contain fossils, petrified wood, salt, silt, clay, minerals, and rock pieces. The type of sedimentary rock that will form depends upon the kind of sediment that it is made of. Shale, the most abundant sedimentary rock, is made of clay and silt. Sandstone has bigger particles that are made from quartz sand.

14   Lots of Science Library Book #9

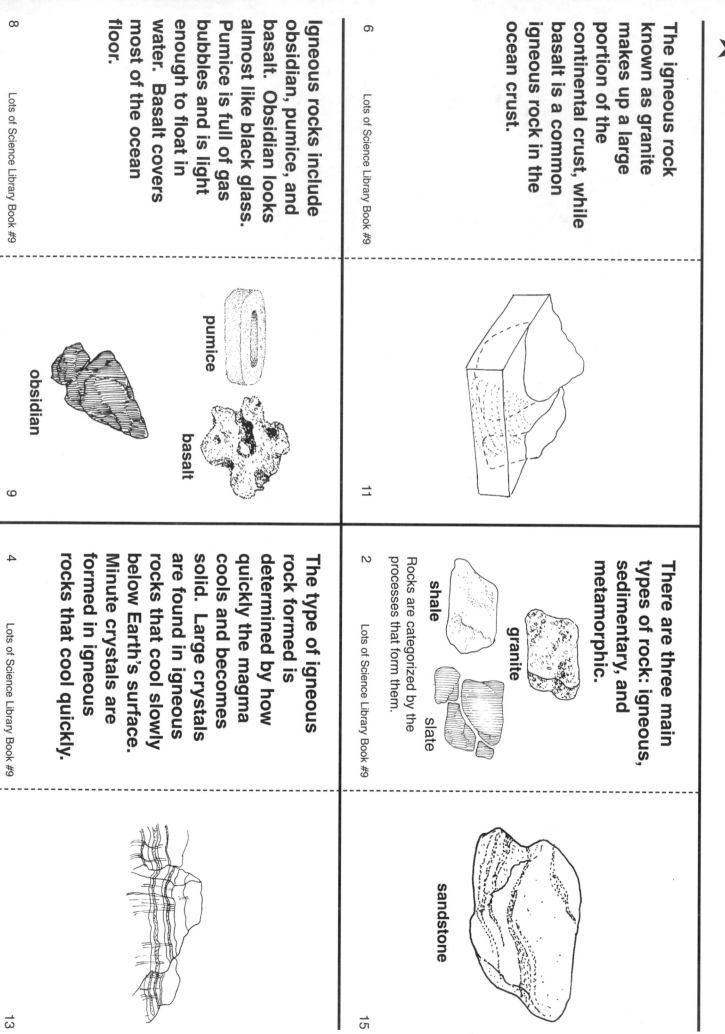

⭐

The igneous rock known as granite makes up a large portion of the continental crust, while basalt is a common igneous rock in the ocean crust.

Igneous rocks include obsidian, pumice, and basalt. Obsidian looks almost like black glass. Pumice is full of gas bubbles and is light enough to float in water. Basalt covers most of the ocean floor.

pumice

basalt

obsidian

There are three main types of rock: igneous, sedimentary, and metamorphic.

shale

granite

slate

Rocks are categorized by the processes that form them.

sandstone

The type of igneous rock formed is determined by how quickly the magma cools and becomes solid. Large crystals are found in igneous rocks that cool slowly below Earth's surface. Minute crystals are formed in igneous rocks that cool quickly.

Sediments from all types of rocks collect in layers to become sedimentary rocks. These break down through weathering to become sediments or undergo heat and pressure to be transformed into metamorphic rocks.

There are three types of rocks: igneous, sedimentary, and metamorphic. In *Lots of Science Library Book #9,* we learned about igneous and sedimentary rocks.

Under extreme heat, all types of rock can melt and become magma.

Rocks in the lithosphere change over time. The rock cycle shows how old rocks become transformed into new ones. Examine the next two pages.

Magma cools to become igneous rocks. Igneous rocks can break down through weathering to become sediments or they can be transformed by heat and pressure into metamorphic rocks.

slate

Metamorphic rocks can either break down through weathering to become sediments or they can be transformed by heat and pressure into other metamorphic rocks.

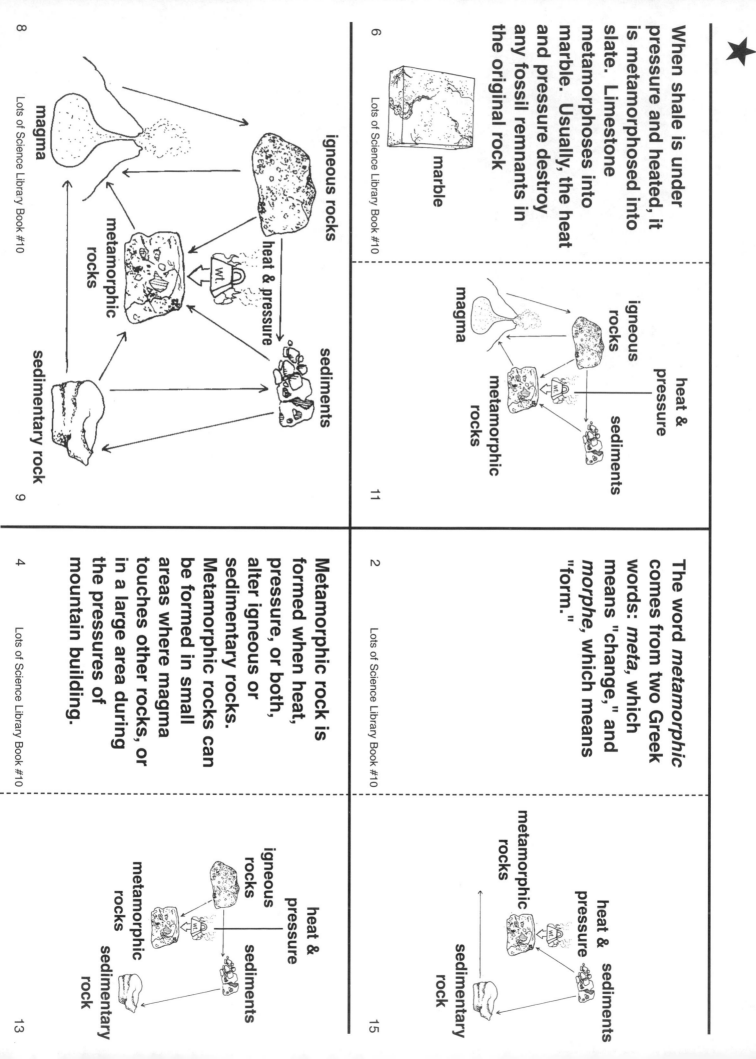

★

When shale is under pressure and heated, it is metamorphosed into slate. Limestone metamorphoses into marble. Usually, the heat and pressure destroy any fossil remnants in the original rock

igneous rocks

heat & pressure

sediments

metamorphic rocks

magma

sedimentary rock

---

marble

---

igneous rocks

heat & pressure

sediments

metamorphic rocks

magma

The word *metamorphic* comes from two Greek words: *meta*, which means "change," and *morphe*, which means "form."

metamorphic rocks

heat & sediments pressure

sedimentary rock

Metamorphic rock is formed when heat, pressure, or both, alter igneous or sedimentary rocks. Metamorphic rocks can be formed in small areas where magma touches other rocks, or in a large area during the pressures of mountain building.

igneous rocks

heat & pressure

sediments

metamorphic rocks

sedimentary rock

gold

Softest
1 Talc
2 Gypsum
3 Calcite
4 Flourite
5 Apatite
6 Feldspar
7 Quartz
8 Topaz
9 Corundum
10 Diamond

Hardest

Compound minerals are combinations of two or more elements. These include sulfides, lead, and antimony.

Minerals can be divided into two groups: native and compound. Native elements are minerals made up of a pure element. These minerals have a chemical symbol for their name. These include gold (Au), silver (Ag), copper (Cu), and carbon (C).

Carbon occurs as a native element in the form of diamonds and graphite.

Minerals are naturally-made, nonliving, crystalline substances made of elements. Minerals make up Earth's rocks and soil. Minerals are also found on Earth's moon and on other rocky planets — Mercury, Venus, and Mars.

Rocks can have any number of minerals in them. Minerals give rocks their color, hardness, texture, density, and luster.

Minerals are used in everyday life. Graphite is found in lead pencils. Gypsum is used in school chalk. Sodium chloride is household salt.

Diamonds are minerals made of pure carbon. This carbon is formed in an igneous rock called kimberlite in the upper mantle of the lithosphere. As the kimberlite rock is subjected to heat and pressure, diamond is formed.

8

Lots of Science Library Book #11

diamond in Kimberlite

9

The variations in the crystal structures of minerals make them very different from one another. The Mohs scale ranks mineral hardness from 1-10, with talc, the softest mineral, ranking as 1 and diamond, the hardest mineral, ranking as 10.

6

Lots of Science Library Book #11

gold

silver

copper

carbon

11

Most minerals develop in liquids such as magma. Magma contains all the kinds of atoms that make up Earth's minerals. As magma cools, crystals of different shapes form, resulting in different kinds of minerals.

4

Lots of Science Library Book #11

lead

antimony

sulfides

13

Scientists have identified nearly 3,000 different minerals in the lithosphere. Minerals include salts, metal ores, and gemstones.

2

Lots of Science Library Book #11

metal ore

salt

emerald in limestone

15

Fascinating Physical Features of Earth

Miners today often have to dig about 2 tons (2 tonnes) of rock to find only 1 ounce (28 grams) of gold. Some of the world's largest gold mines are in South Africa.

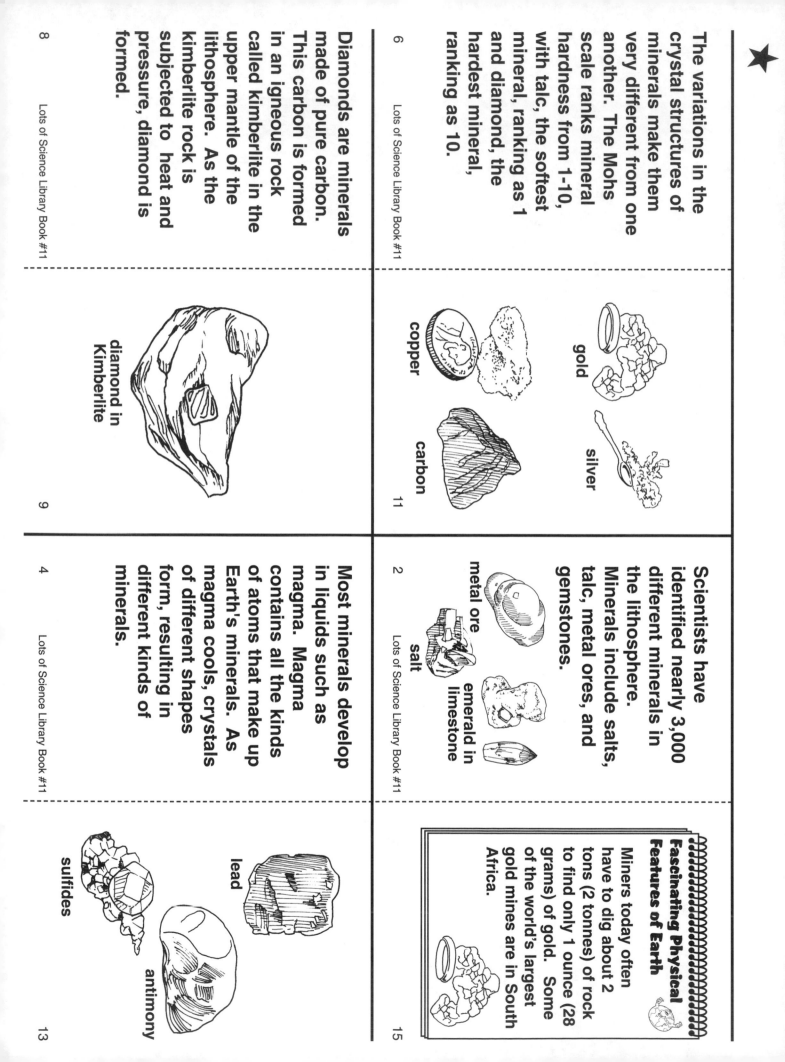

For example, when a skeleton is covered by sediment, it decays slowly. As mineral-rich water passes through and around the bones, it slowly replaces the decaying cells with mineral deposits.

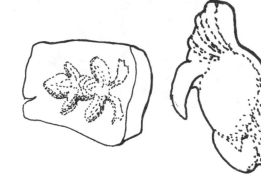

Wood can also be fossilized, or transformed into stone. If wood is buried in an area where hot, silica-rich water flows, it can become petrified when minerals fill the empty cells of the decaying wood.

A mold fossil is the preserved imprint of a plant or animal that has not been filled in with minerals or sediments. Mold fossils are often found in limestone, and are often seen in sedimentary building stones.

Fossils are the preserved remains of plants and animals that were once alive.

In rare cases, animal tissue has been preserved when it has been buried in hot, silica-rich water, frozen in ice, mummified in deserts, or immersed in tar pits.

Since the cells of petrified wood are replaced slowly, sometimes cell by cell, many details of the original wood can be seen in the fossil — rings, knots, scars, and bark.

The hard parts of an organism are often preserved as fossils—horns, teeth, bones, claws, scales, and shells.

6

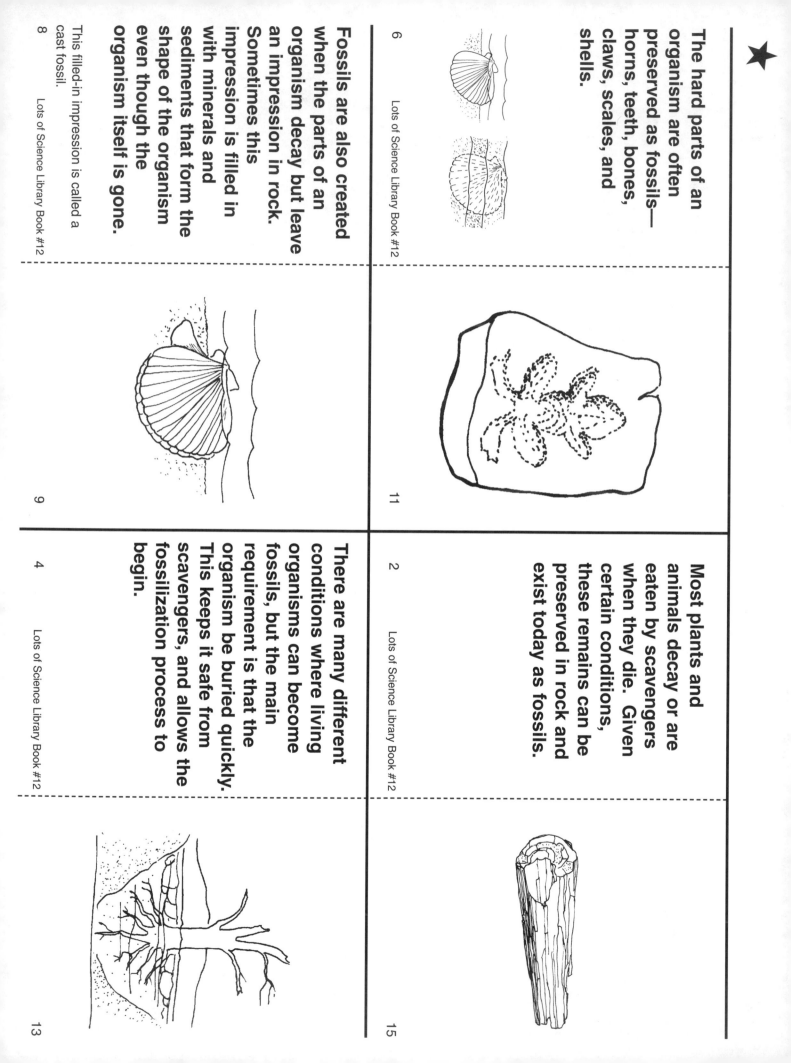

8

This filled-in impression is called a cast fossil.

Fossils are also created when the parts of an organism decay but leave an impression in rock. Sometimes this impression is filled in with minerals and sediments that form the shape of the organism even though the organism itself is gone.

9

11

Most plants and animals decay or are eaten by scavengers when they die. Given certain conditions, these remains can be preserved in rock and exist today as fossils.

2

There are many different conditions where living organisms can become fossils, but the main requirement is that the organism be buried quickly. This keeps it safe from scavengers, and allows the fossilization process to begin.

4

13

15

Five main factors affect soil--the parent rocks that are weathered, organic materials, climate, time, and the geography of the land.

5

Some rocks contain minerals that resist weathering. Soil from these rocks will develop slowly and contain course grains.

7

Climate can be the most important factor in the formation of soil. High temperatures can speed up chemical reactions. Low temperatures can inhibit plant growth, limiting the amount of organic material in soil.

Creatures such as ants, earthworms, and termites stir up the soil, mixing the organic materials with the other layers of sediment and water.

Earthworms are important in this process, since they not only mix the soil, but also change its texture and chemical nature.

Soil is the layer of weathered rock and minerals that covers Earth's land. Its depth varies from less than an inch thick to hundreds of feet thick.

mountains

ocean

ocean crust

continental crust

mantle

1

Without soil, there would be few plants and very little food for the animals on Earth.

3

Fascinating Physical Features of Earth

In Java, pineapples and winged beans are grown in alternate rows to keep the soil fertile. If the same crop is grown over and over in the same soil, the nutrients get used up and are not replaced.

16

Soil on a steep slope is seldom deep and is easily eroded. On level ground, soil may be deep, but it can be eroded by wind or it may get waterlogged.

14

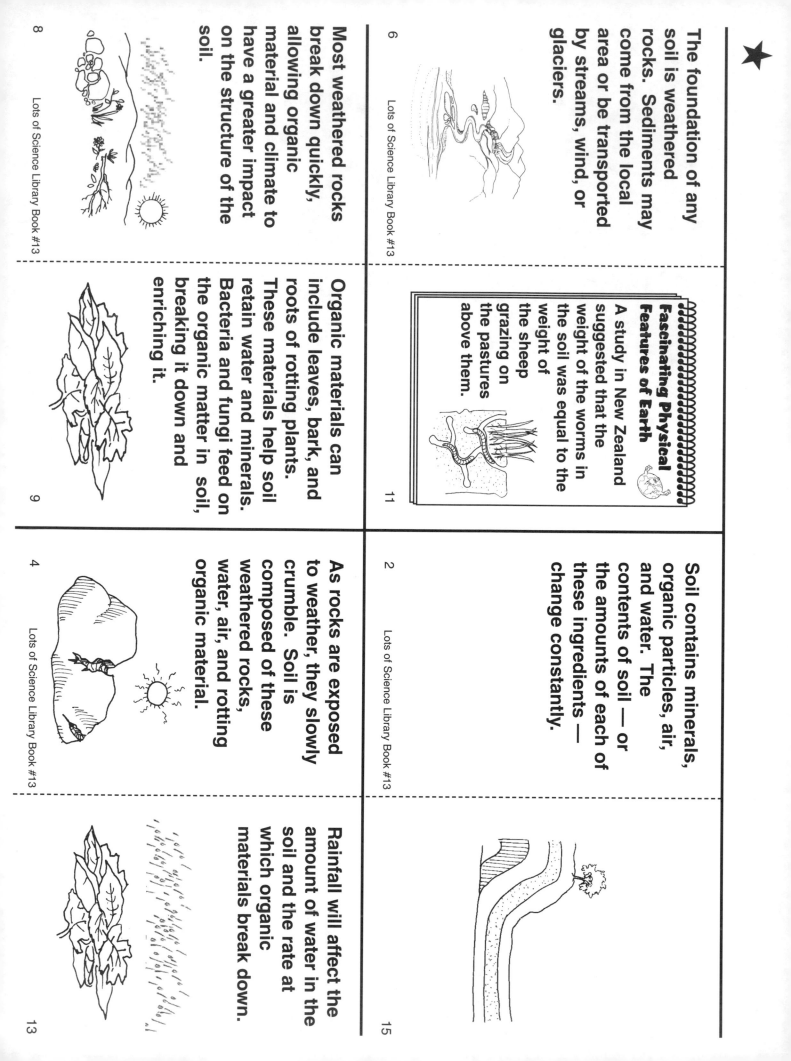

6 — The foundation of any soil is weathered rocks. Sediments may come from the local area or be transported by streams, wind, or glaciers.

8 — Most weathered rocks break down quickly, allowing organic material and climate to have a greater impact on the structure of the soil.

11 — Fascinating Physical Features of Earth

A study in New Zealand suggested that the weight of the worms in the soil was equal to the weight of the sheep grazing on the pastures above them.

9 — Organic materials can include leaves, bark, and roots of rotting plants. These materials help soil retain water and minerals. Bacteria and fungi feed on the organic matter in soil, breaking it down and enriching it.

2 — Soil contains minerals, organic particles, air, and water. The contents of soil — or the amounts of each of these ingredients — change constantly.

4 — As rocks are exposed to weather, they slowly crumble. Soil is composed of these weathered rocks, water, air, and rotting organic material.

15 — Rainfall will affect the amount of water in the soil and the rate at which organic materials break down.

13

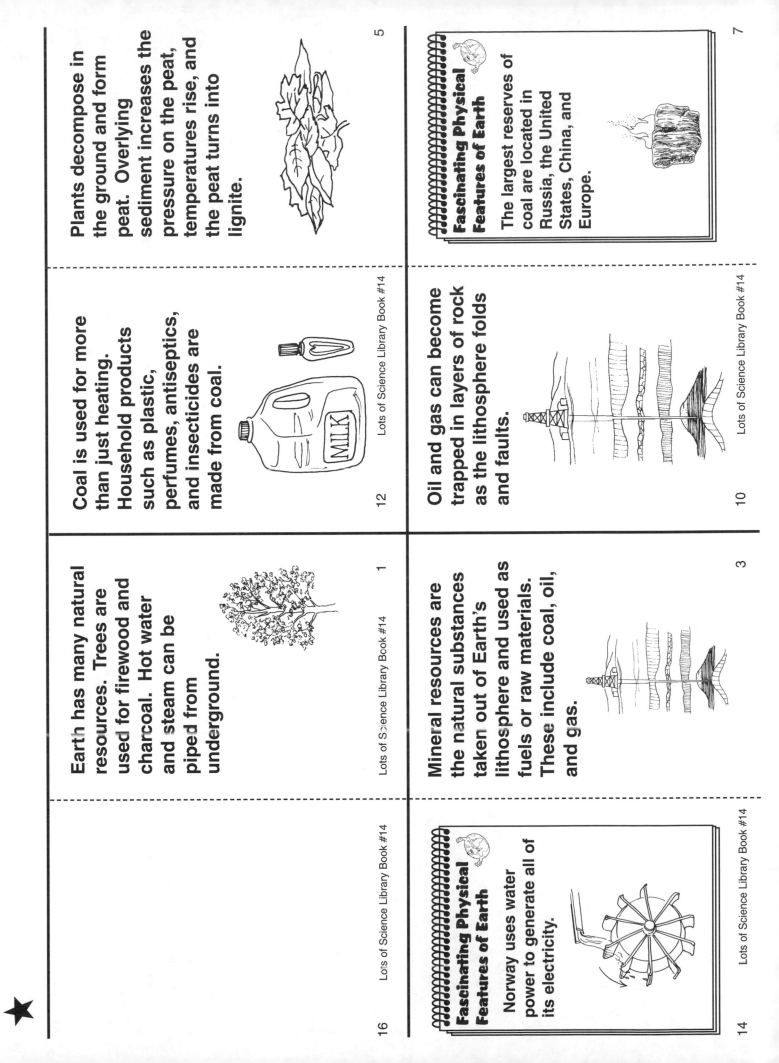

5

Plants decompose in the ground and form peat. Overlying sediment increases the pressure on the peat, temperatures rise, and the peat turns into lignite.

7

Fascinating Physical Features of Earth

The largest reserves of coal are located in Russia, the United States, China, and Europe.

Coal is used for more than just heating. Household products such as plastic, perfumes, antiseptics, and insecticides are made from coal.

Oil and gas can become trapped in layers of rock as the lithosphere folds and faults.

Earth has many natural resources. Trees are used for firewood and charcoal. Hot water and steam can be piped from underground.

Mineral resources are the natural substances taken out of Earth's lithosphere and used as fuels or raw materials. These include coal, oil, and gas.

3

Fascinating Physical Features of Earth

Norway uses water power to generate all of its electricity.

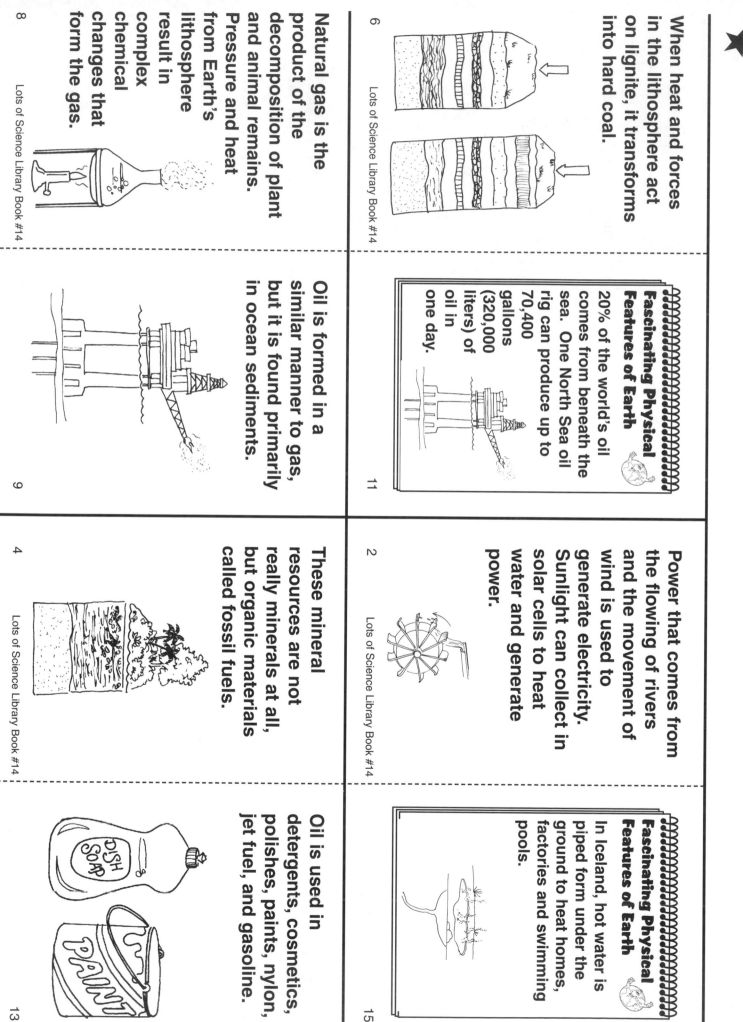

Pressure and heat from Earth's lithosphere result in complex chemical changes that form the gas.

When heat and forces in the lithosphere act on lignite, it transforms into hard coal.

★

Natural gas is the product of the decomposition of plant and animal remains.

9

Oil is formed in a similar manner to gas, but it is found primarily in ocean sediments.

11

Fascinating Physical Features of Earth

20% of the world's oil comes from beneath the sea. One North Sea oil rig can produce up to 70,400 gallons (320,000 liters) of oil in one day.

Power that comes from the flowing of rivers and the movement of wind is used to generate electricity. Sunlight can collect in solar cells to heat water and generate power.

These mineral resources are not really minerals at all, but organic materials called fossil fuels.

Oil is used in detergents, cosmetics, polishes, paints, nylon, jet fuel, and gasoline.

DISH SOAP

PAINT

13

Fascinating Physical Features of Earth

In Iceland, hot water is piped form under the ground to heat homes, factories and swimming pools.

15

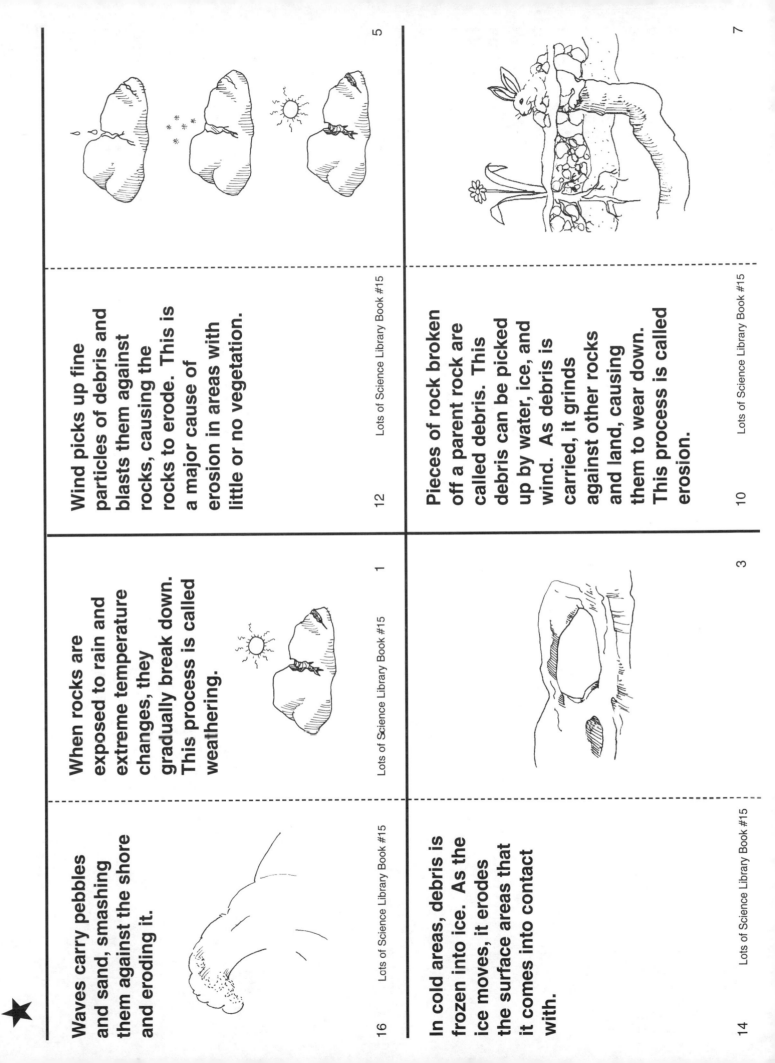

Wind picks up fine particles of debris and blasts them against rocks, causing the rocks to erode. This is a major cause of erosion in areas with little or no vegetation.

Pieces of rock broken off a parent rock are called debris. This debris can be picked up by water, ice, and wind. As debris is carried, it grinds against other rocks and land, causing them to wear down. This process is called erosion.

When rocks are exposed to rain and extreme temperature changes, they gradually break down. This process is called weathering.

Waves carry pebbles and sand, smashing them against the shore and eroding it.

In cold areas, debris is frozen into ice. As the ice moves, it erodes the surface areas that it comes into contact with.

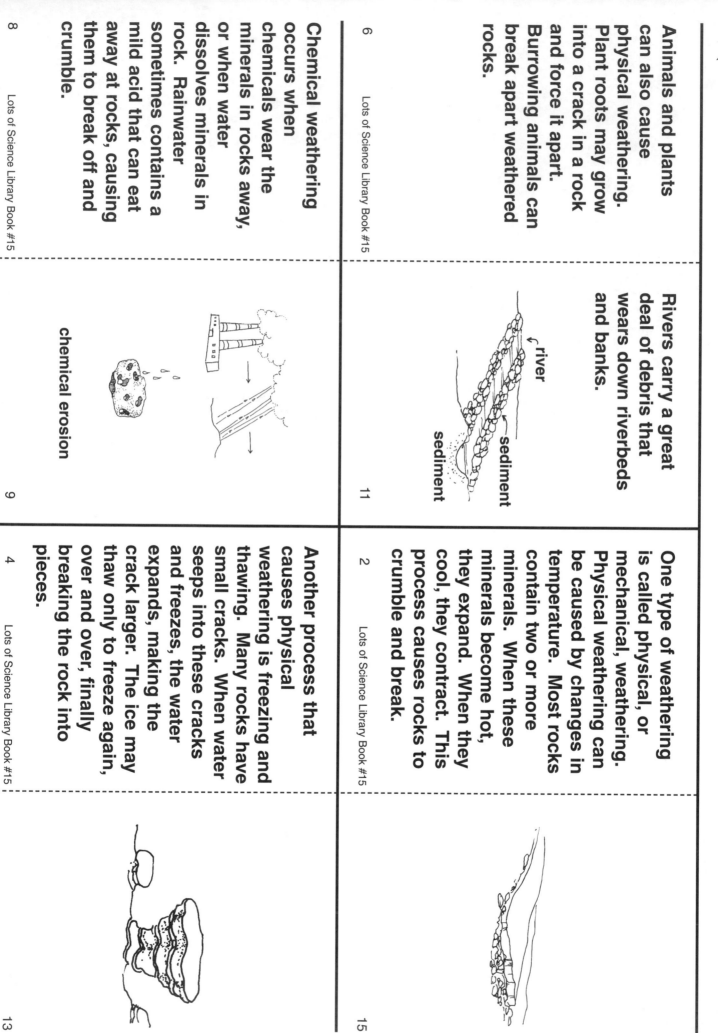

★

Animals and plants can also cause physical weathering. Plant roots may grow into a crack in a rock and force it apart. Burrowing animals can break apart weathered rocks.

8

Chemical weathering occurs when chemicals wear the minerals in rocks away, or when water dissolves minerals in rock. Rainwater sometimes contains a mild acid that can eat away at rocks, causing them to break off and crumble.

chemical erosion

9

6

Rivers carry a great deal of debris that wears down riverbeds and banks.

← river

sediment

sediment

11

One type of weathering is called physical, or mechanical, weathering. Physical weathering can be caused by changes in temperature. Most rocks contain two or more minerals. When these minerals become hot, they expand. When they cool, they contract. This process causes rocks to crumble and break.

2

Another process that causes physical weathering is freezing and thawing. Many rocks have small cracks. When water seeps into these cracks and freezes, the water expands, making the crack larger. The ice may thaw only to freeze again, over and over, finally breaking the rock into pieces.

4

15

13

Fascinating Physical Features of Earth

The longest canyon in the world is the Grand Canyon on the Colorado River in Arizona. It is 277 miles (446 km) long.

After the glacier is gone, the granite bedrock is left exposed, and without the pressure and weight of the glacier on it, the granite will bulge upward. Manhattan Island is an example of exposed granite bedrock.

Complete mountain ranges can erode away, leaving only a few rock formations standing. These standing pieces are called monadnocks. Stone Mountain in Georgia is an example of a monadnock.

A mesa is a broad, flat-topped hill rising above the surrounding land. A butte is similar to a mesa with a smaller, narrower top and steep sides. These features are seen in the southwestern United States.

Some of the most dramatic land features on Earth are the result of erosion.

Many unusual natural sculptures are created by erosion.

wind blown sand

sand

Erosion shapes nearly every feature of Earth's surface, large or small. This natural bridge was formed when the rocks beneath it eroded away.

Sometimes a glacier will erode most of the sedimentary rock in an area.

Sometimes water erodes land, leaving rocky places that did not erode intact. These remaining high areas are not true mountains, although they are higher than the surrounding land. A mesa is an example of such a feature.

Streams and rivers can carve deep canyons in land through erosion. These canyons are often the best places to observe layers of sedimentary rocks.

Fascinating Physical Features of Earth

The deepest canyon in the world is Colca Canyon in Peru. It is over 10,500 feet (3,223 m) deep.

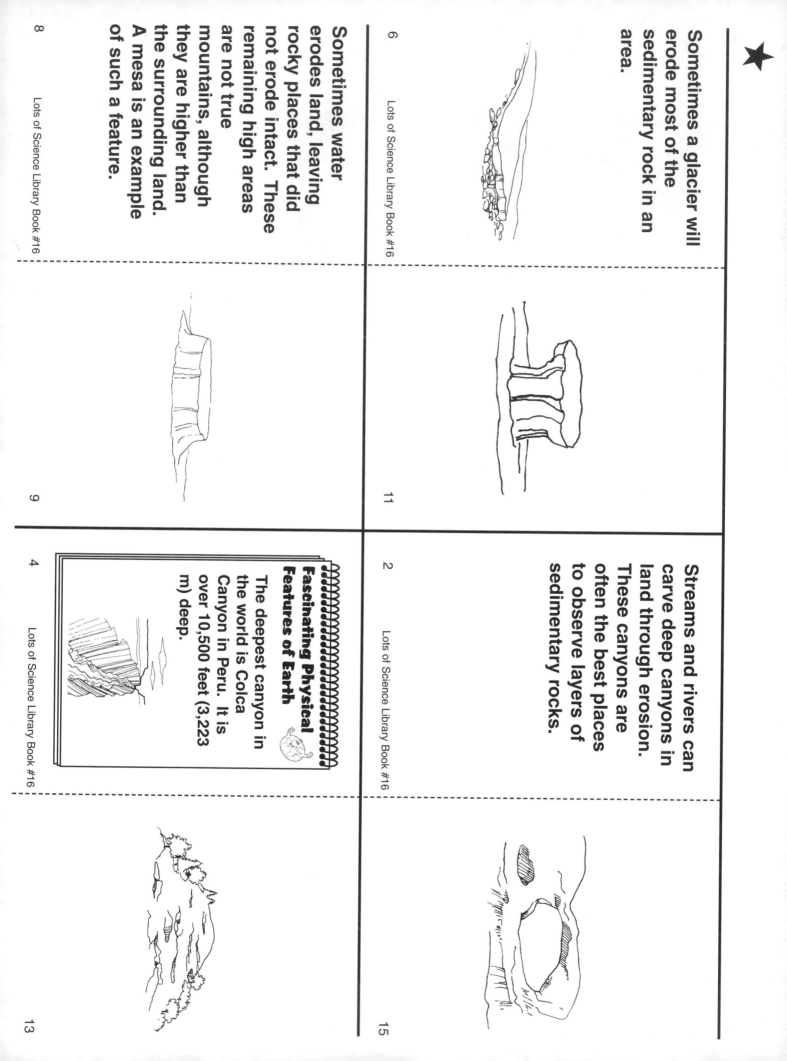

underground water passageway

stream

cave mouth

It is easy to remember the names of these formations. Stalactites have to hang on "tite" to stay in place.

Ground water dissolves the mineral calcite (calcium carbonate) from limestone. When the mineral rich water enters the cave and evaporates, the calcite is left behind. The deposited calcite slowly forms stalactites and stalagmites.

Erosion takes place above and below the surface of Earth. Underground water erodes and weathers rock in the same way that surface water does.

stream

impermeable rock

impermeable rock

permeable limestone

Fascinating Physical Features of Earth

The longest cave system is the Mammoth Cave National Park in Kentucky. It is 348 miles (484 km) long.

Sometimes s stalactite and a stalagmite will meet, forming a pillar, or cave column.

pillar

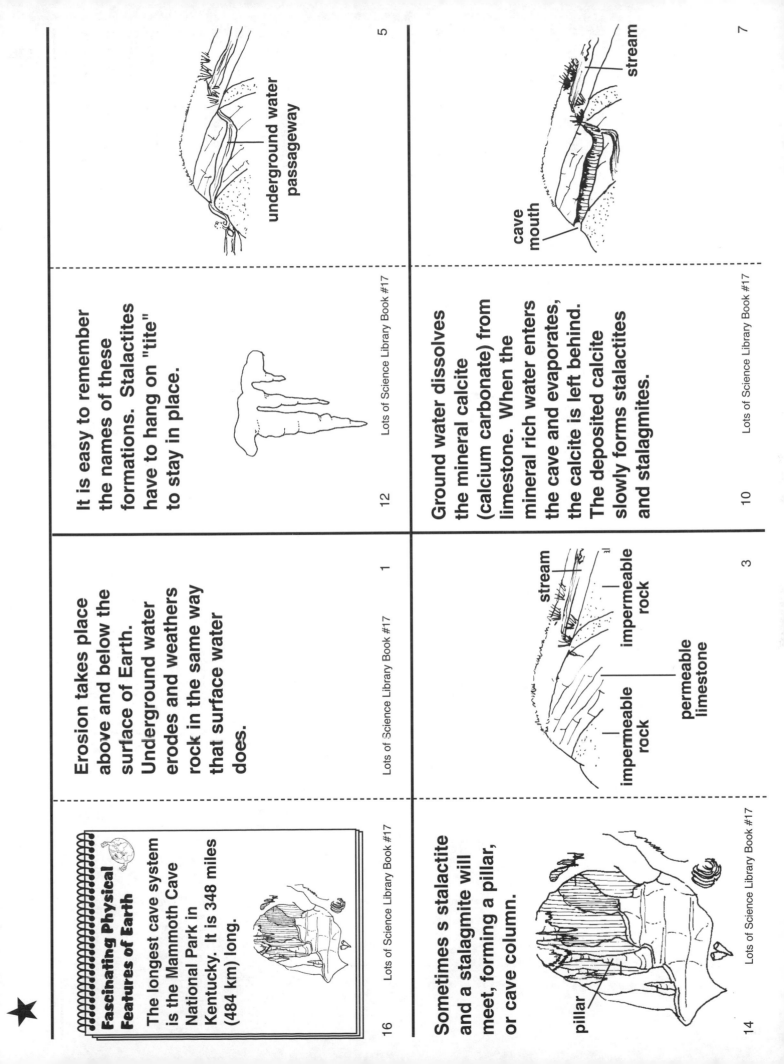

These openings become larger over time as more water moves through the passageways. Rocks along the passageways weather away and collapse, only to be eroded by the moving water.

Caves are formed under ground when moving water erodes soft rock. Caves are most commonly found in limestone, although they can form in other rocks.

Many rocks contain minerals that dissolve easily in water. As water runs through rock passageways, it collects these minerals. When the water evaporates, the minerals are left behind.

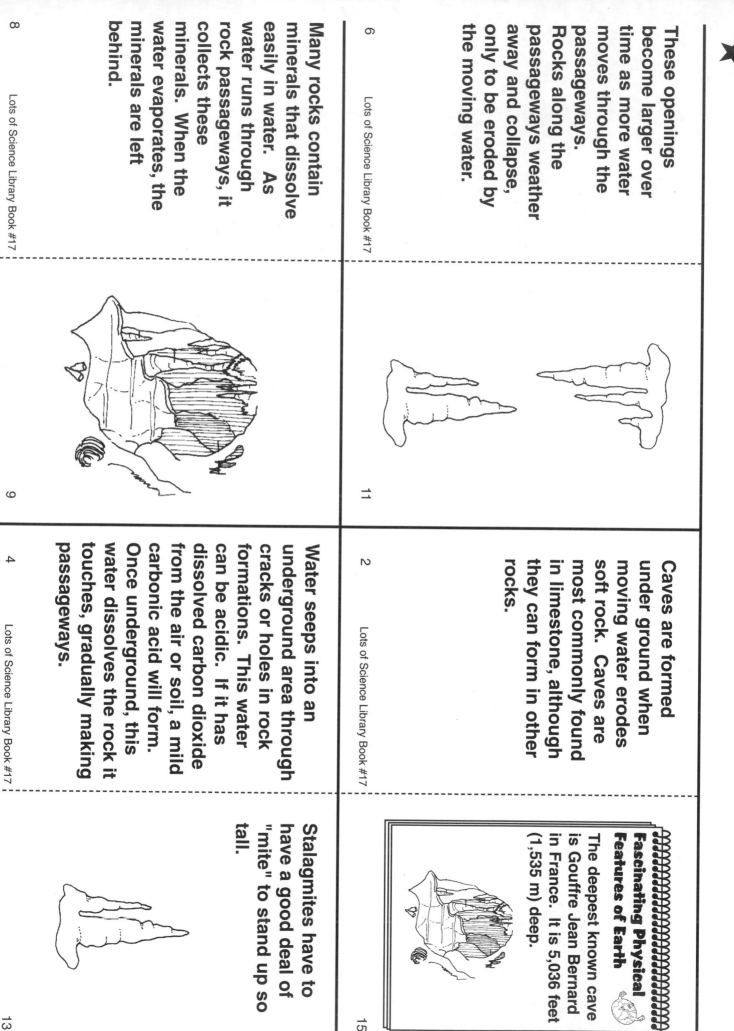

Water seeps into an underground area through cracks or holes in rock formations. This water can be acidic. If it has dissolved carbon dioxide from the air or soil, a mild carbonic acid will form. Once underground, this water dissolves the rock it touches, gradually making passageways.

Stalagmites have to have a good deal of "mite" to stand up so tall.

Fascinating Physical Features of Earth

The deepest known cave is Gouffre Jean Bernard in France. It is 5,036 feet (1,535 m) deep.

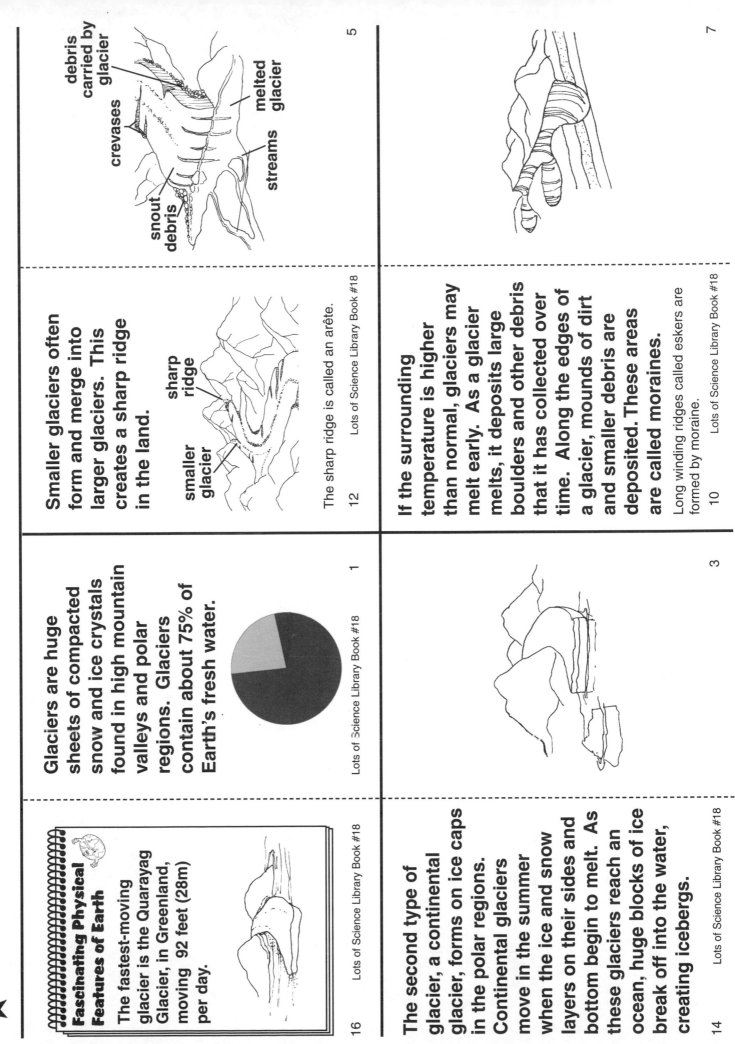

debris carried by glacier

crevases

snout debris

melted glacier

streams

5

Smaller glaciers often form and merge into larger glaciers. This creates a sharp ridge in the land.

sharp ridge

smaller glacier

The sharp ridge is called an arête.

12   Lots of Science Library Book #18

Glaciers are huge sheets of compacted snow and ice crystals found in high mountain valleys and polar regions. Glaciers contain about 75% of Earth's fresh water.

1

Lots of Science Library Book #18

Fascinating Physical Features of Earth

The fastest-moving glacier is the Quarayag Glacier, in Greenland, moving 92 feet (28m) per day.

16   Lots of Science Library Book #18

The second type of glacier, a continental glacier, forms on ice caps in the polar regions. Continental glaciers move in the summer when the ice and snow layers on their sides and bottom begin to melt. As these glaciers reach an ocean, huge blocks of ice break off into the water, creating icebergs.

14   Lots of Science Library Book #18

7

If the surrounding temperature is higher than normal, glaciers may melt early. As a glacier melts, it deposits large boulders and other debris that it has collected over time. Along the edges of a glacier, mounds of dirt and smaller debris are deposited. These areas are called moraines.

Long winding ridges called eskers are formed by moraine.

10   Lots of Science Library Book #18

3

★

Glaciers move from a few inches to a few feet a day, depending on the slope of the mountain, the amount of snow that falls, and the size of the glacier. As a glacier moves, it erodes the land, creating a U-shaped valley with steep sides and a flat floor.

8

The front of the glacier is called the snout.

6

Sometimes a small lake is left behind in the glacier valley.

11

As moving glaciers reach warmer lower altitudes, they melt, causing streams of water to move ahead of them.

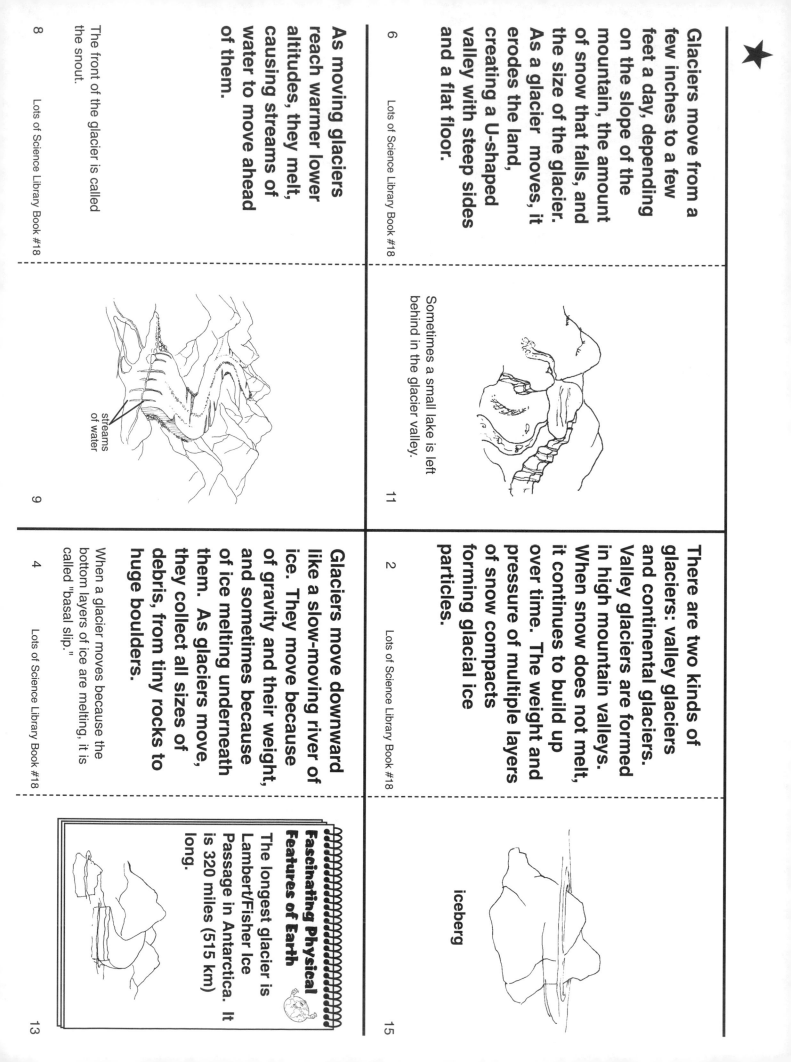

streams of water

9

There are two kinds of glaciers: valley glaciers and continental glaciers. Valley glaciers are formed in high mountain valleys. When snow does not melt, it continues to build up over time. The weight and pressure of multiple layers of snow compacts forming glacial ice particles.

2

iceberg

15

Glaciers move downward like a slow-moving river of ice. They move because of gravity and their weight, and sometimes because of ice melting underneath them. As glaciers move, they collect all sizes of debris, from tiny rocks to huge boulders.

When a glacier moves because the bottom layers of ice are melting, it is called "basal slip."

4

Fascinating Physical Features of Earth

The longest glacier is Lambert/Fisher Ice Passage in Antarctica. It is 320 miles (515 km) long.

13

permeable rock

source

rain

river

Finer particles of clay and sand stay suspended in the water. Minerals from the eroded rocks are carried by the river and deposited downstream.

river

sediment

Water flow may increase if other rivers feed into the main flow. Small streams or rivers that flow into a larger river are called tributaries.

The point where the tributary meets the main river is called a confluence.

The middle stage begins when the downhill movement is not as steep. The water's speed decreases, and the river may make bends as it flows. These bends are called meanders.

Rivers shape Earth's land. Their movement erodes rocks and deposits large amounts of sediment along the way.

river

sediment

sediment

sediment

As a river erodes the land, it forms banks that contain the water.

river banks

river banks

river banks

Fascinating Physical Features of Earth

The second longest river is the Amazon in South America at 4,000 miles long (6,400 km long).

The lower stage of the river flows into an ocean or a lake. It deposits much of its sediment at this stage.

lower stage

As the water of a river flows, it carries sediment and debris. Heavier rocks and pebbles are rolled and bounced along the riverbed, causing them to become rounded and smooth.

This process is called attrition.

A large river has three stages: the upper stage, middle stage, and lower stage.

middle stage

lower stage

upper stage

The amount of erosion that takes place depends on the amount of water in a river, the speed of the water, and the composition of the rocks it flows over. Some rocks, such as sandstone, erode quickly, while granite is more resistant to erosion.

meanders

The upper stage tends to be a V-shaped valley with steep sides formed by fast-flowing water moving downhill.

The beginning of a river is called the source. Rivers usually start in hills or mountains. Their water comes from rain, melting snow, or underground springs. Most rivers travel to an ocean or a lake.

tributaries

Fascinating Physical Features of Earth

The world's longest river is the Nile in North Africa at 4,160 miles long (6,695 km long).

Fascinating Physical Features of Earth

The largest waterfall in the world is Angel Falls in Venezuela at 3,212 feet high (979 m high).

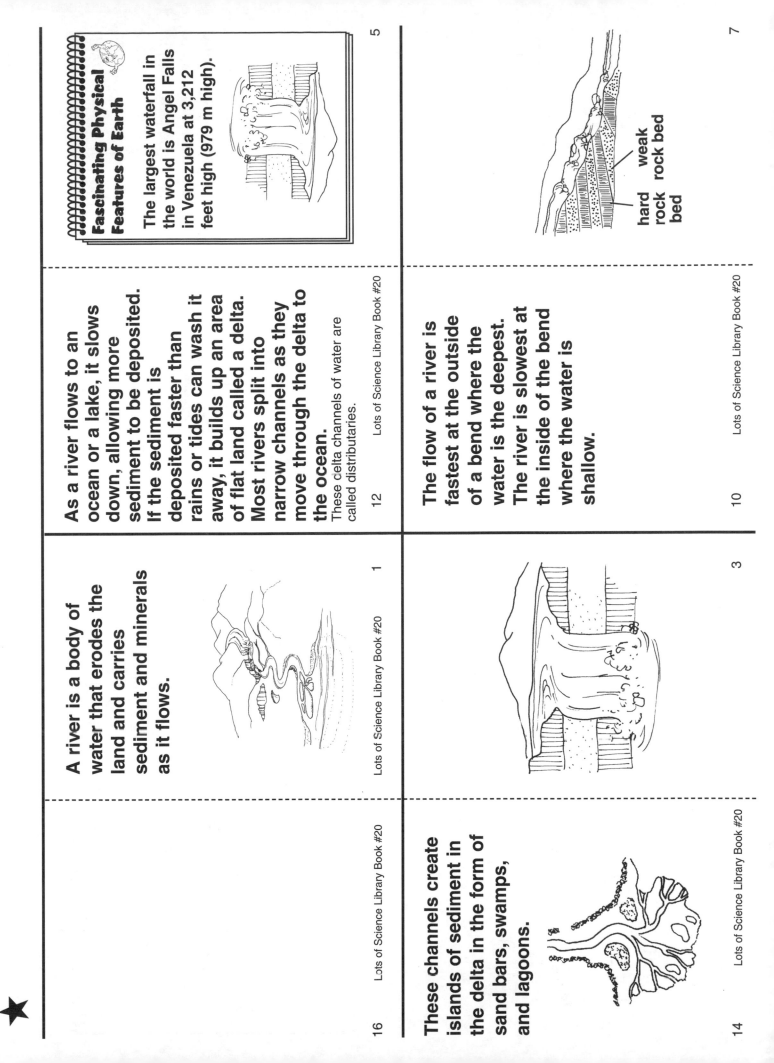

weak rock bed

hard rock bed bed

As a river flows to an ocean or a lake, it slows down, allowing more sediment to be deposited. If the sediment is deposited faster than rains or tides can wash it away, it builds up an area of flat land called a delta. Most rivers split into narrow channels as they move through the delta to the ocean.

These delta channels of water are called distributaries.

The flow of a river is fastest at the outside of a bend where the water is the deepest. The river is slowest at the inside of the bend where the water is shallow.

A river is a body of water that erodes the land and carries sediment and minerals as it flows.

These channels create islands of sediment in the form of sand bars, swamps, and lagoons.

8

As rivers flow down slightly sloped areas, they tend to wind backward and forward in a snake-like manner. As rivers near an ocean or a lake, they often form horseshoe shaped bends called meanders.

★

Sometimes an area will have soft rock and hard rock deposited close to one another or in layers. As the soft rock is eroded, the hard rock stays in place, creating rapids.

---

meanders

---

water is deepest and fastest

As rivers flow down slightly sloped areas, they tend to wind backward and forward in a snake-like manner. As rivers near an ocean or a lake, they often form horseshoe shaped bends called meanders.

Fascinating Physical Features of Earth

Iguazu Falls in Brazil are 2 1/2 miles (4km) wide and 260 ft (80m) high. During the rainy season, the volume of water pouring over the falls every second could fill about 6 Olympic-size swimming pools.

As a river flows, it may travel from a hard to a weak rock bed. Over time, the weaker rock layers will erode away until resistant rock is exposed. This process results in a waterfall. The land formation under a waterfall resembles a steep hill that was formed by the erosion of the soft rock.

Fascinating Physical Features of Earth

In Bangladesh, millions of people live on the islands formed in the delta of the River Ganges. They grow rice and raise cattle.

rift valley lake

5) Artificial lakes are created when people dig a hollow area in the land and fill it with water. Diverted river water can be used to fill a lake.

---

glacier lake

4) Oxbow lakes are formed when the meander of a river is cut off from the rest of the river.

A lake is a body of water surrounded by land. Natural lakes are created when a large amount of water is collected in a hollow area of hard, or impermeable, rock.

Lakes do not have a long life span. They are easily altered by weather and sediment deposits from rivers and streams that feed into them. These deposits make lake bottoms very muddy.

Fascinating Physical Features of Earth

The deepest lake in the world is Lake Baikal in Russia. It is about 12,000 square miles (30,912 square km) and 5,315 – 6,365 feet (1,620 – 1,940 m) deep. It holds so much water that all five of the Great Lakes could be emptied into this lake bed.

★

2) A glacier lake is created when the land is eroded by the weight of a glacier and then filled in as ice melts.

oxbow lake

Lake water comes from rivers, flooding, melting glaciers, and underground water traveling downhill.

3) Crater lakes are formed when volcano craters are filled with water.

crater lake

river

There are five main ways that a lake can be formed. **1)** The deepest and most stable lakes are called rift valley lakes. They are formed when shifts in Earth's plates form depressions that fill up with water.

Sometimes dams are built on a river to control water flow and to flood a flat area, creating a lake.

artificial lake

Fascinating Physical Features of Earth

The largest lake is Lake Superior at 31,700 square miles (82,100 square km).

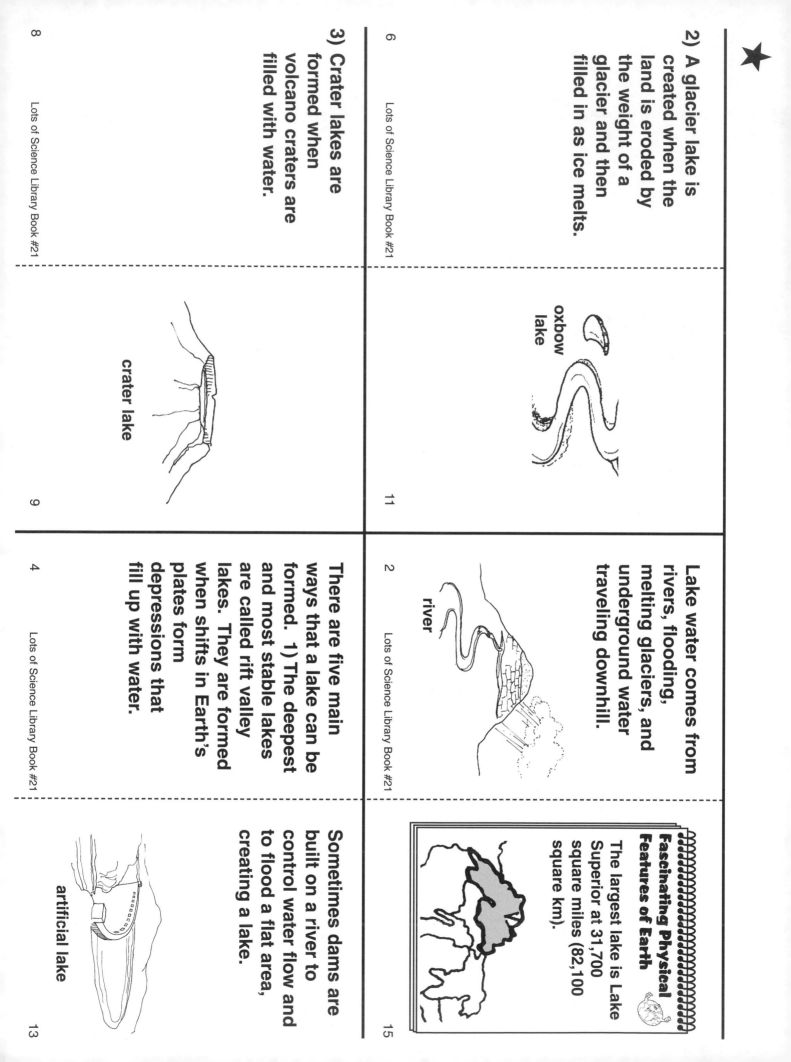

greatest
amount
of
erosion

One such formation is called an arch. It is created as a wall of rock is eroded by ocean waves. If the top of the arch collapses, then a stack is formed.

If the rock forming shore line cliffs contains cracks, air is squeezed into the cracks as waves hit the rock. When the waves retreat, the air pressure is released and air pushes out again, making the cracks larger over time. Eventually, this process will break down existing rocks and create new shoreline formations.

This process is called hydraulic action.

Coastlines are some of the most rapidly changing landscapes on Earth. Wind, rain, and the movement of ocean water erodes coastlines.

The water that is carried onto a shore by tides or waves contains tiny pieces of rock and shell. This sediment in the water increases the amount of erosion that takes place on the shore.

Fascinating Physical Features of Earth

If all the coastlines on Earth were straightened out, they would stretch nearly 13 times around Earth's equator.

Fascinating Physical Features of Earth

The sand dunes on the Atlantic coast of France reach heights of 300 feet (91 m) and move inland at about 20 feet (6 m) per year.

Waves usually hit a shore at an angle and then retreat in a straight line. This causes debris to be deposited on a shore in a zigzag manner. This process is called longshore drift.

Waves are created by wind traveling over water. The size of a wave depends on the speed of the wind and how long the wind has been blowing over the water. Waves cause a great deal of shore erosion. As waves near a shore, they pick up debris from the ocean floor and carry it to the shore.

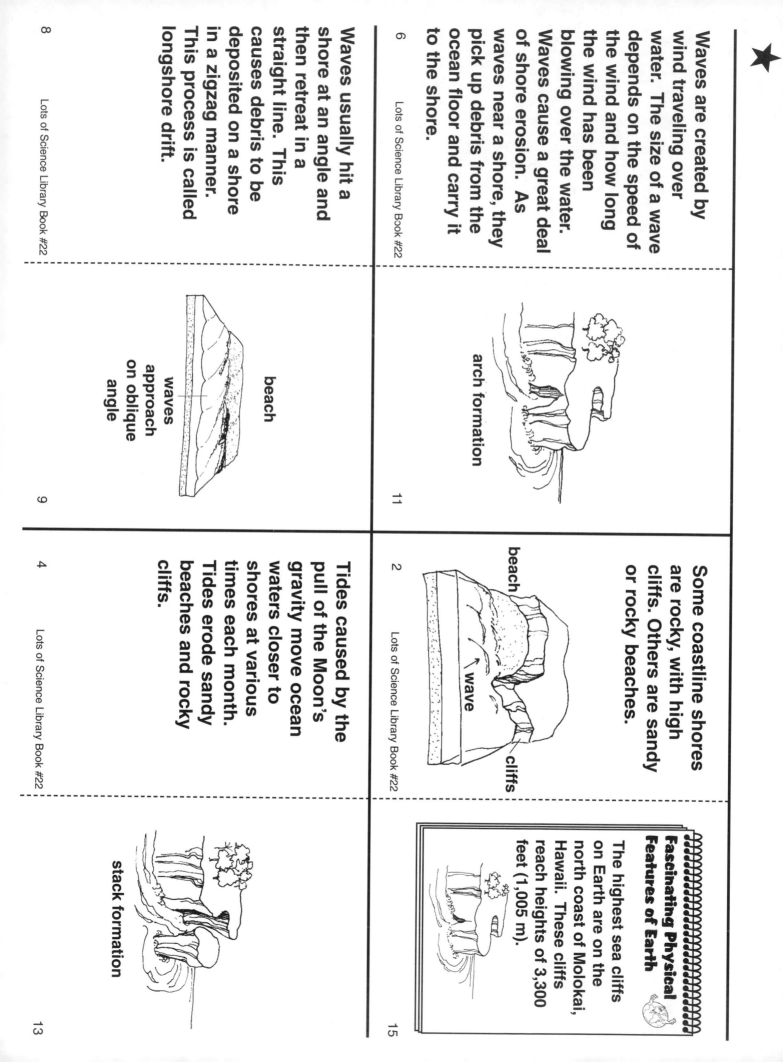

beach

waves approach on oblique angle

arch formation

Tides caused by the pull of the Moon's gravity move ocean waters closer to shores at various times each month. Tides erode sandy beaches and rocky cliffs.

beach

wave

cliffs

Some coastline shores are rocky, with high cliffs. Others are sandy or rocky beaches.

Fascinating Physical Features of Earth

The highest sea cliffs on Earth are on the north coast of Molokai, Hawaii. These cliffs reach heights of 3,300 feet (1,005 m).

stack formation

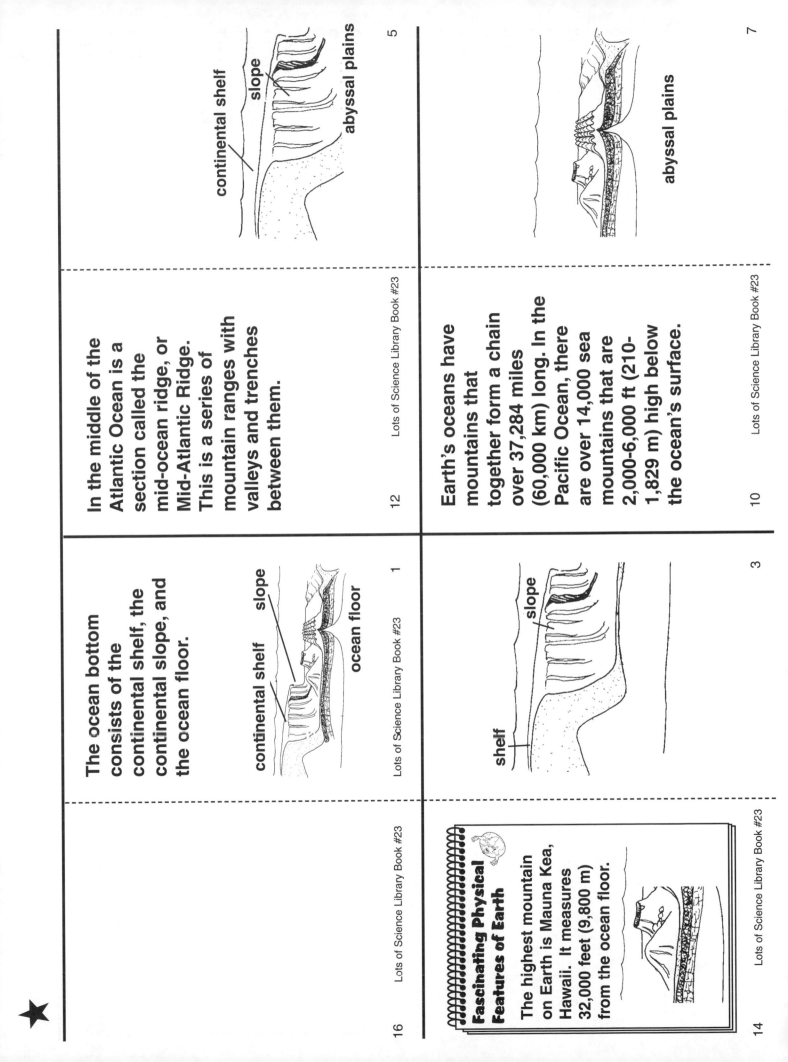

continental shelf
slope
abyssal plains

5

abyssal plains

7

In the middle of the Atlantic Ocean is a section called the mid-ocean ridge, or Mid-Atlantic Ridge. This is a series of mountain ranges with valleys and trenches between them.

12

Earth's oceans have mountains that together form a chain over 37,284 miles (60,000 km) long. In the Pacific Ocean, there are over 14,000 sea mountains that are 2,000-6,000 ft (210-1,829 m) high below the ocean's surface.

10

The ocean bottom consists of the continental shelf, the continental slope, and the ocean floor.

continental shelf   slope

ocean floor

1

shelf
slope

3

16

Fascinating Physical Features of Earth

The highest mountain on Earth is Mauna Kea, Hawaii. It measures 32,000 feet (9,800 m) from the ocean floor.

14

The ocean floor and its abyssal plains contain layers of sand, gravel, and clay. This underwater seabed is a landscape of sandy plains, mountains, plateaus, valleys, and deep trenches.

6

Thousands of volcanoes can be found rising up from the ocean floor. Most of these are the result of the ocean floor spreading. Some rising up to the ocean's surface create volcanic islands.

8

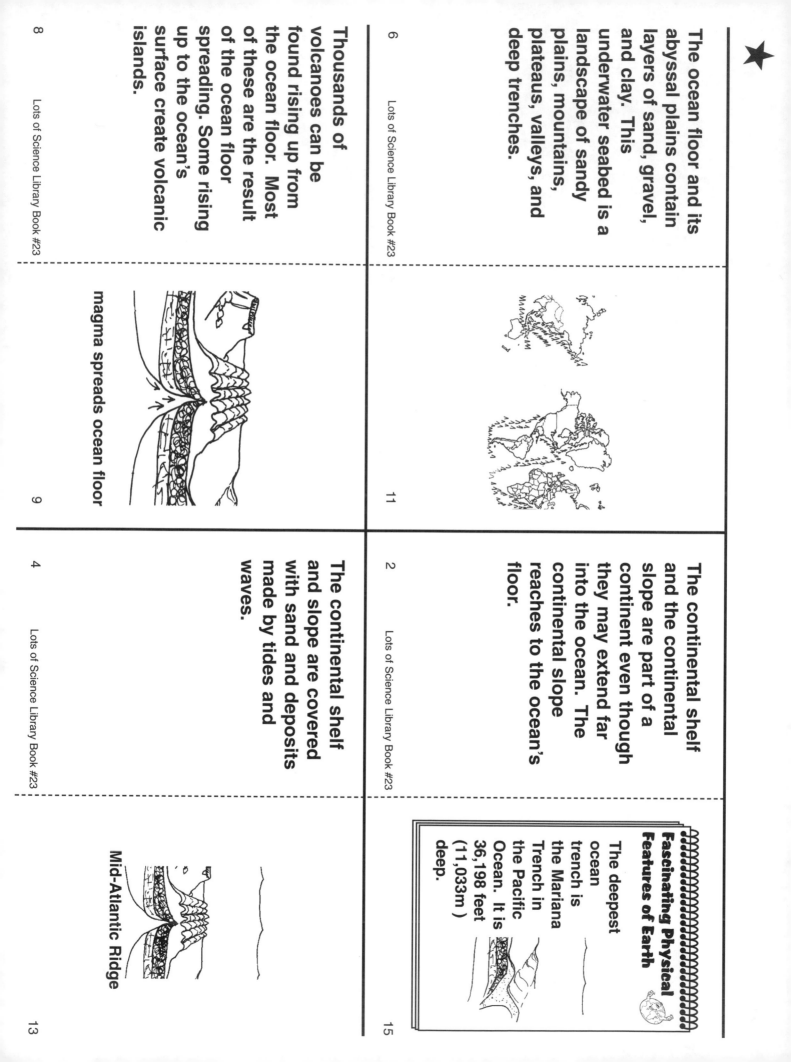

magma spreads ocean floor

9

11

The continental shelf and the continental slope are part of a continent even though they may extend far into the ocean. The continental slope reaches to the ocean's floor.

2

The continental shelf and slope are covered with sand and deposits made by tides and waves.

4

Mid-Atlantic Ridge

13

Fascinating Physical Features of Earth

The deepest ocean trench is the Mariana Trench in the Pacific Ocean. It is 36,198 feet (11,033m) deep.

15

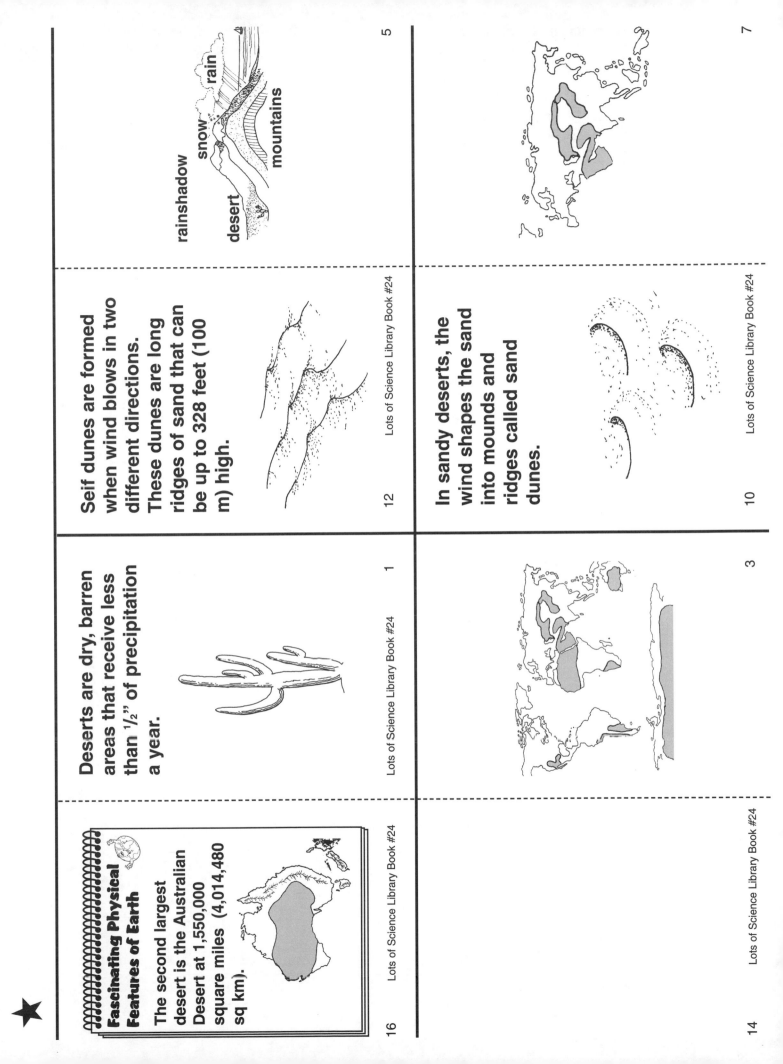

rainshadow

snow

rain

mountains

desert

Seif dunes are formed when wind blows in two different directions. These dunes are long ridges of sand that can be up to 328 feet (100 m) high.

In sandy deserts, the wind shapes the sand into mounds and ridges called sand dunes.

Deserts are dry, barren areas that receive less than 1/2" of precipitation a year.

3

Fascinating Physical Features of Earth

The second largest desert is the Australian Desert at 1,550,000 square miles (4,014,480 sq km).

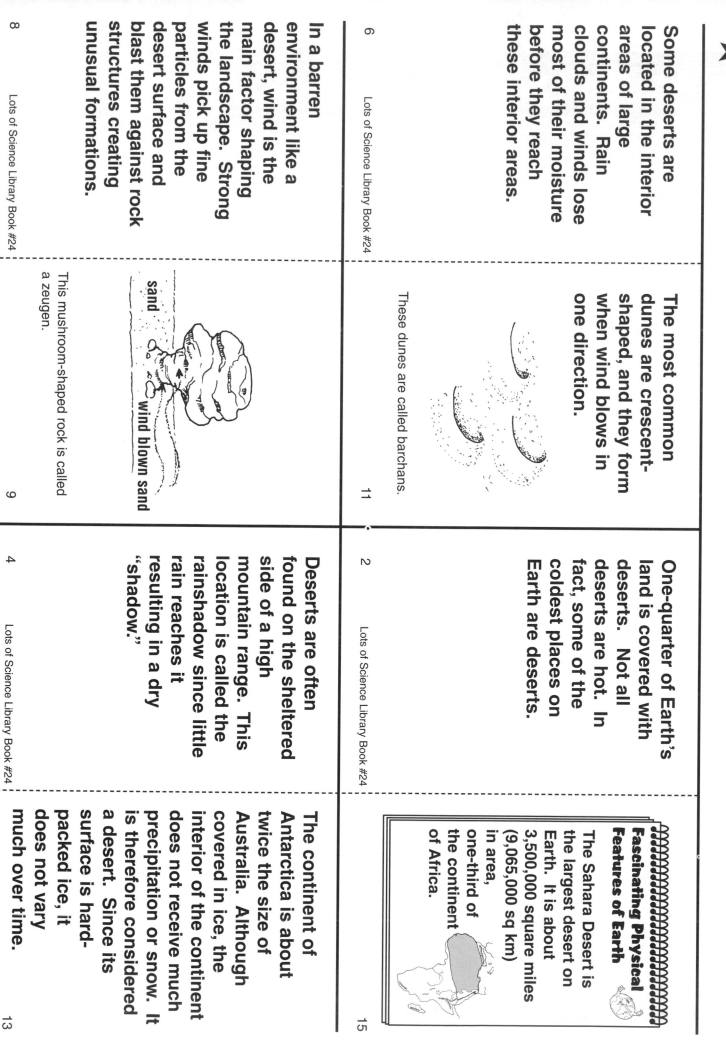

Some deserts are located in the interior areas of large continents. Rain clouds and winds lose most of their moisture before they reach these interior areas.

6

In a barren environment like a desert, wind is the main factor shaping the landscape. Strong winds pick up fine particles from the desert surface and blast them against rock structures creating unusual formations.

8

This mushroom-shaped rock is called a zeugen.

sand

wind blown sand

9

The most common dunes are crescent-shaped, and they form when wind blows in one direction.

These dunes are called barchans.

11

One-quarter of Earth's land is covered with deserts. Not all deserts are hot. In fact, some of the coldest places on Earth are deserts.

2

Deserts are often found on the sheltered side of a high mountain range. This location is called the rainshadow since little rain reaches it resulting in a dry "shadow."

4

The continent of Antarctica is about twice the size of Australia. Although covered in ice, the interior of the continent does not receive much precipitation or snow. It is therefore considered a desert. Since its surface is hard-packed ice, it does not vary much over time.

13

Fascinating Physical Features of Earth

The Sahara Desert is the largest desert on Earth. It is about 3,500,000 square miles (9,065,000 sq km) in area, one-third of the continent of Africa.

15

Great Science Adventures

Graphics Pages

Note: The owner of this book has permission to photocopy the *Graphics Pages* for classroom use only.

Investigative Loop ™

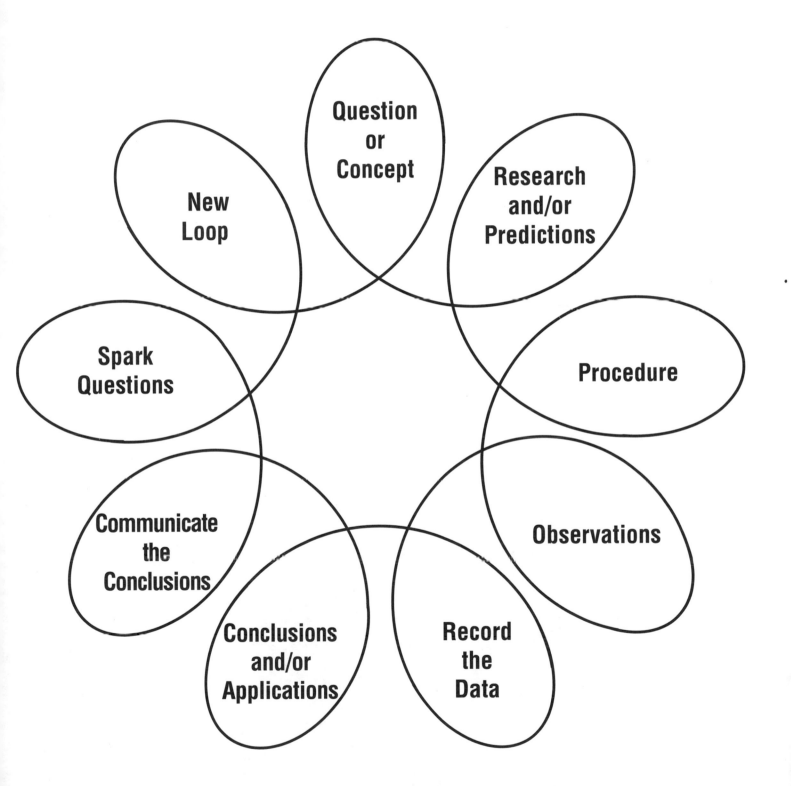

Question or Concept

Research and/or Predictions

New Loop

Procedure

Spark Questions

Observations

Communicate the Conclusions

Record the Data

Conclusions and/or Applications

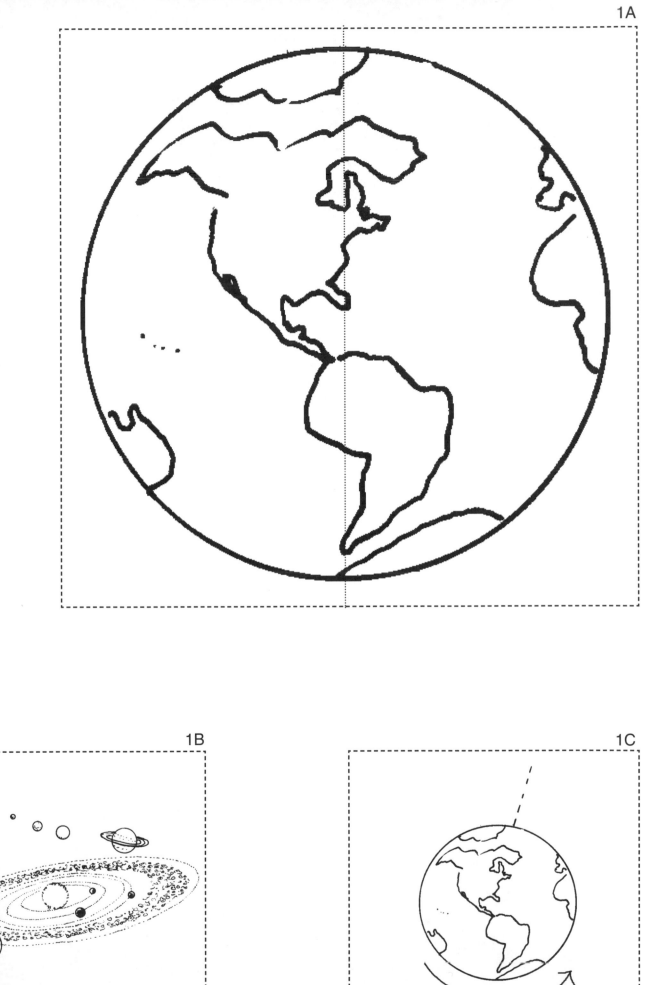

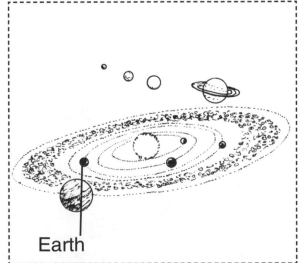

Earth

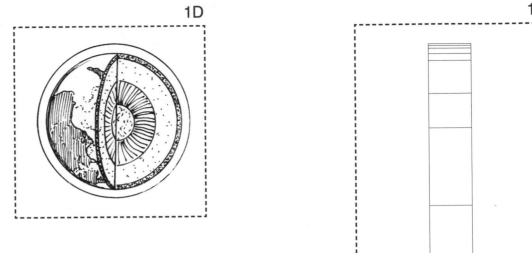

2A
green

2B
blue

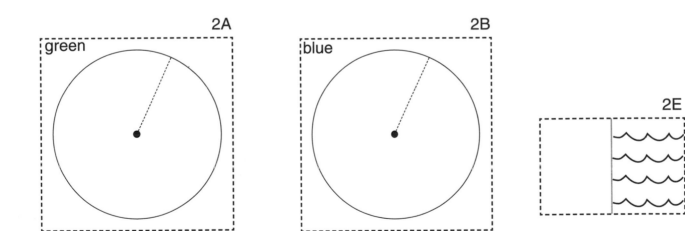

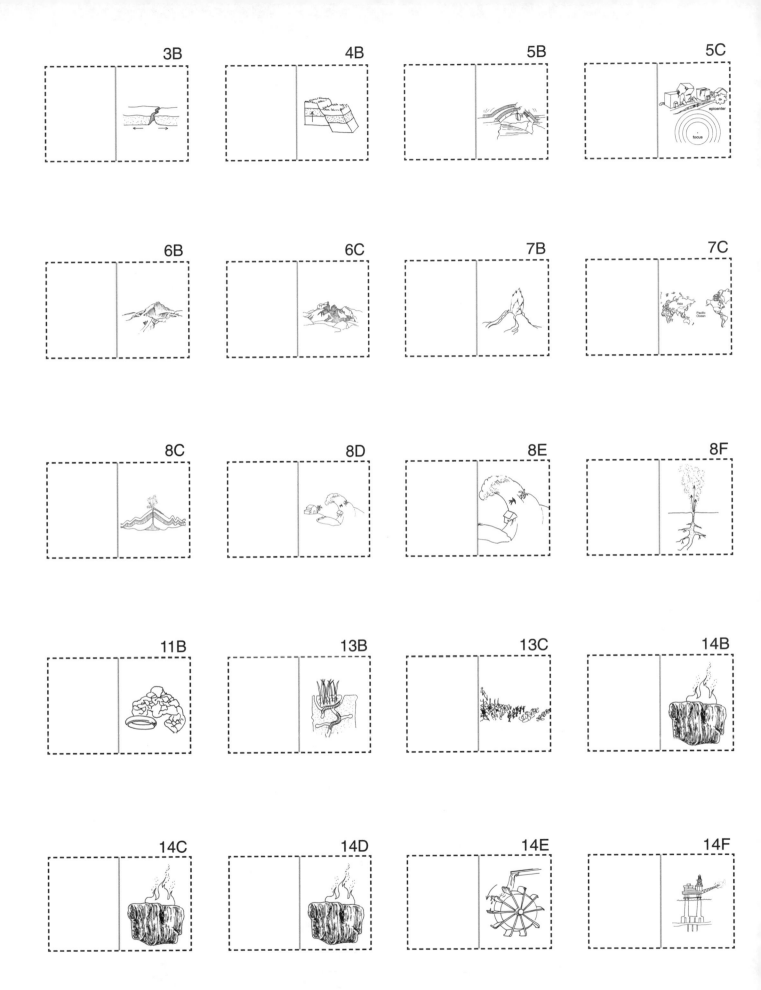

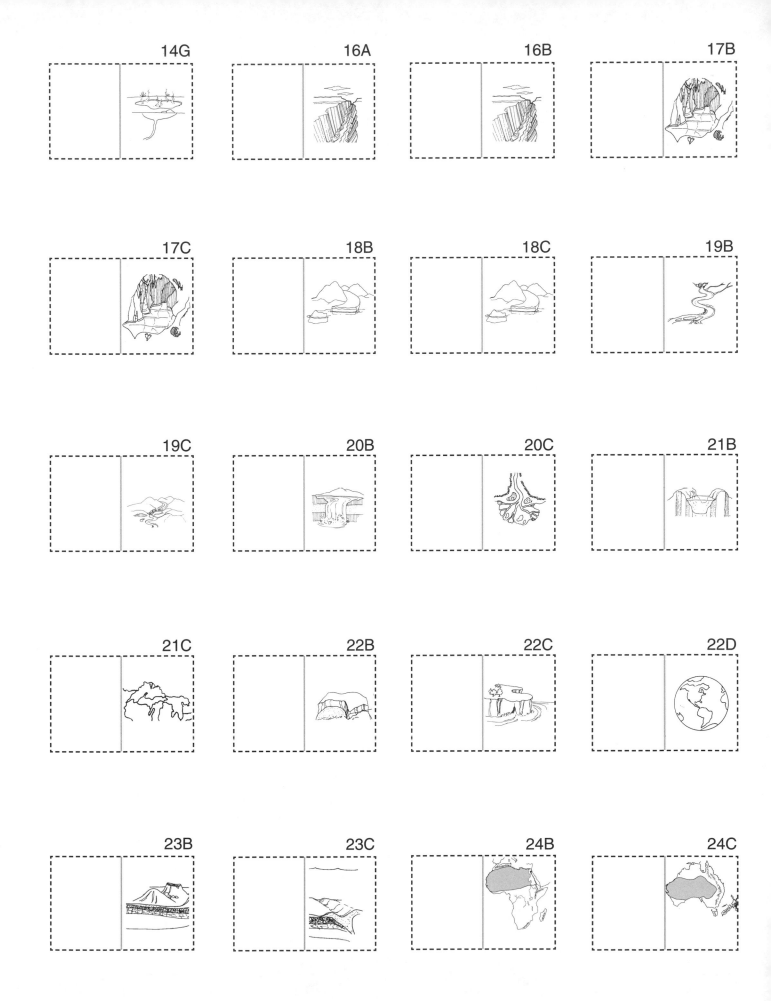

Lab Graphic 13-1

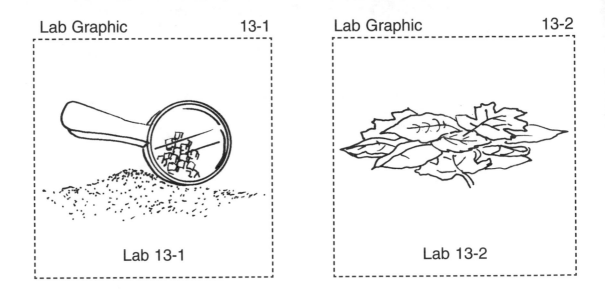

Lab 13-1

Lab Graphic 13-2

Lab 13-2

Lab Graphic 15-1

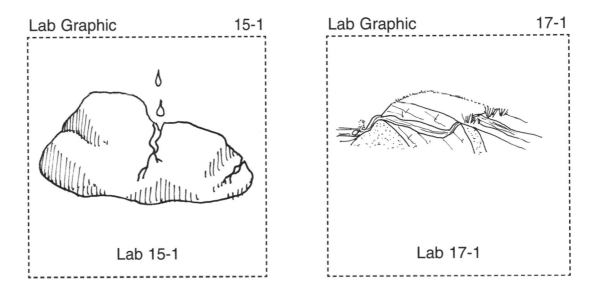

Lab 15-1

Lab Graphic 17-1

Lab 17-1

Lab Graphic 18-1

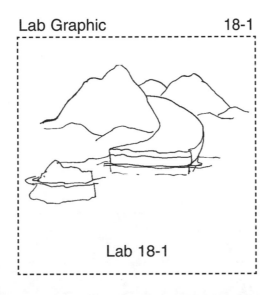

Lab 18-1

glue

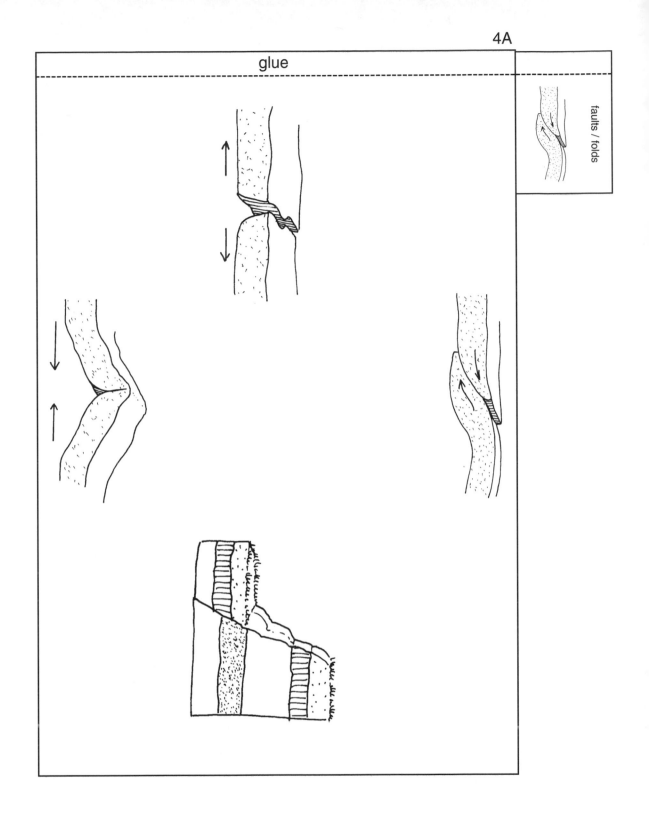

glue

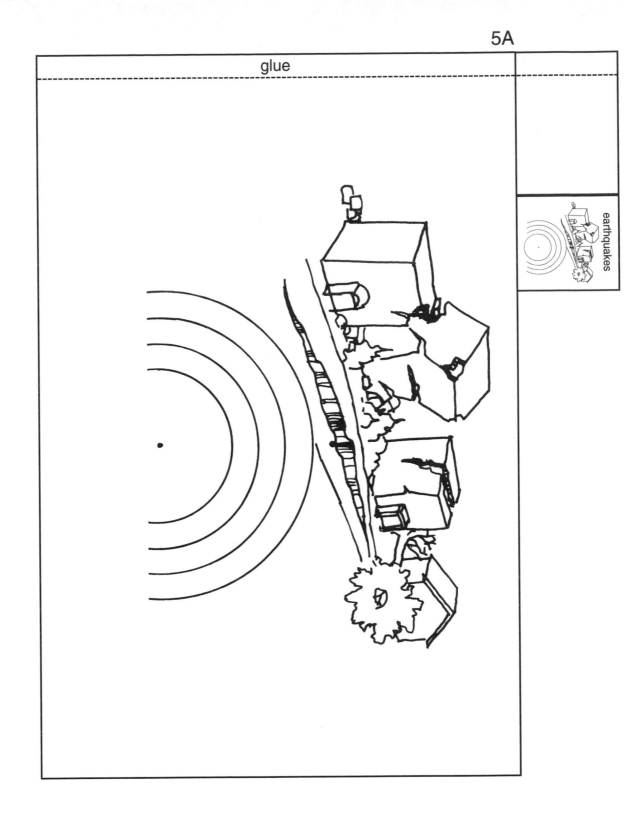

earthquakes

glue

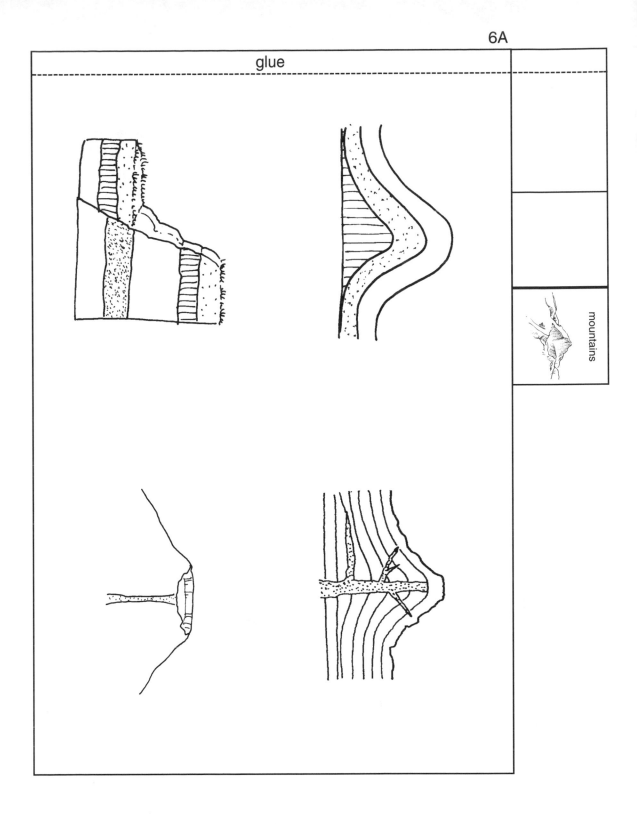

mountains

glue

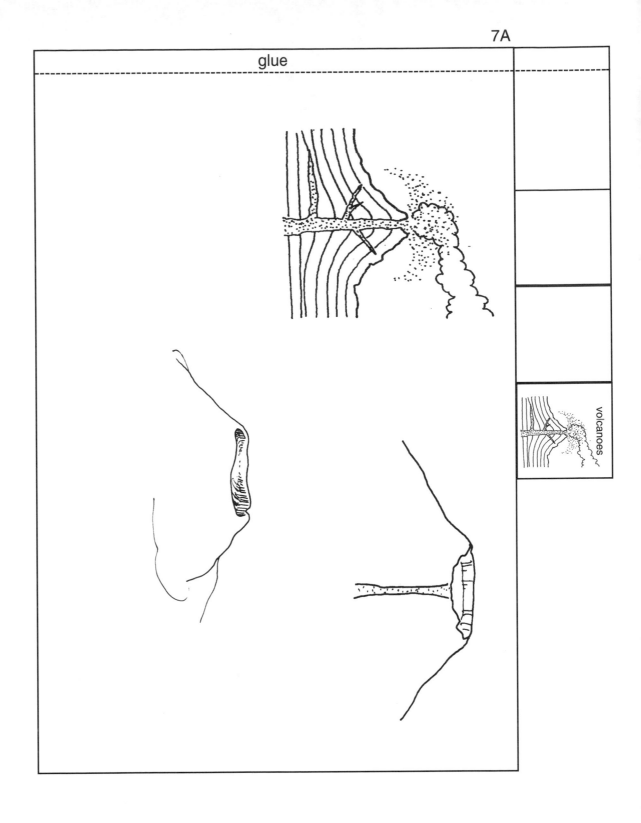

volcanoes

glue

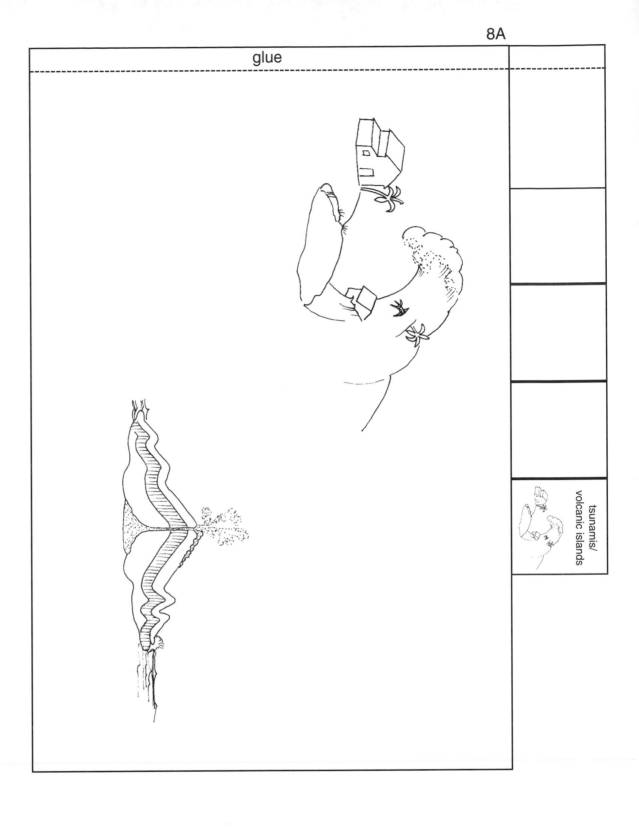

tsunamis/
volcanic islands

glue

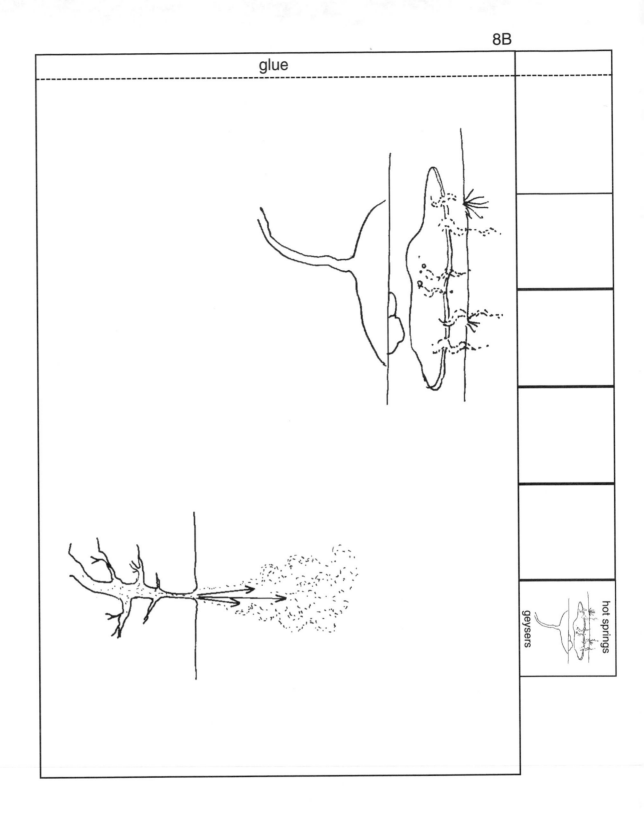

geysers

hot springs

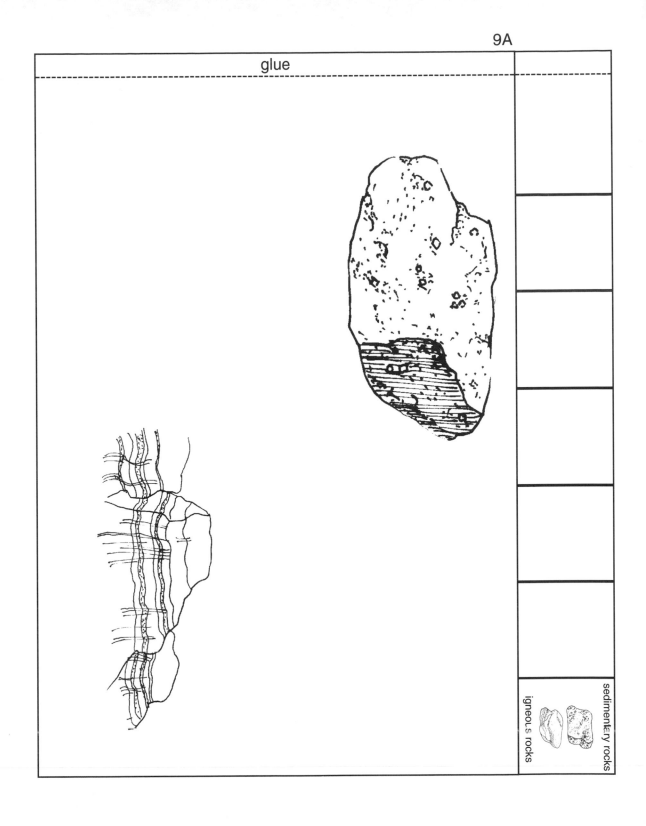

glue

sedimentary rocks

igneous rocks

glue

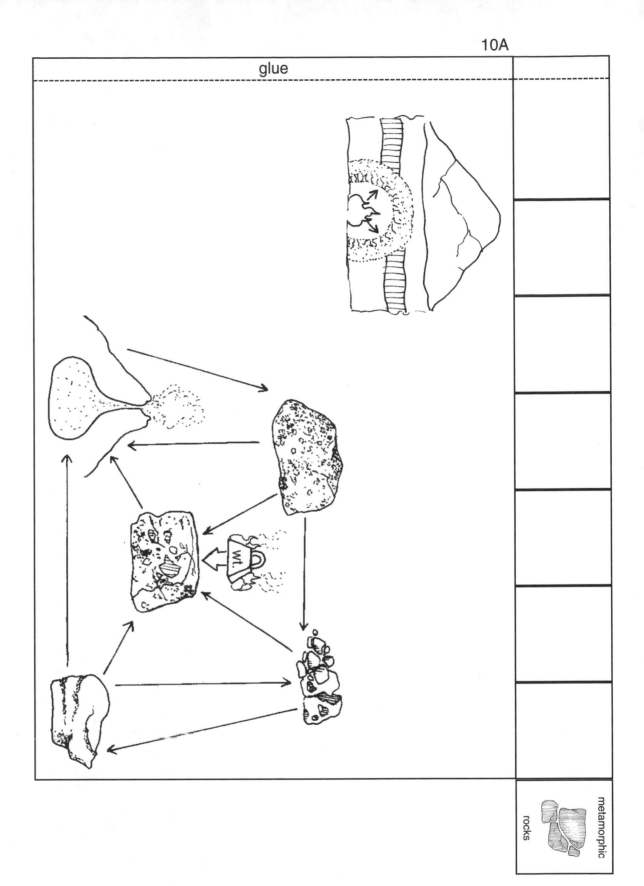

metamorphic rocks

glue

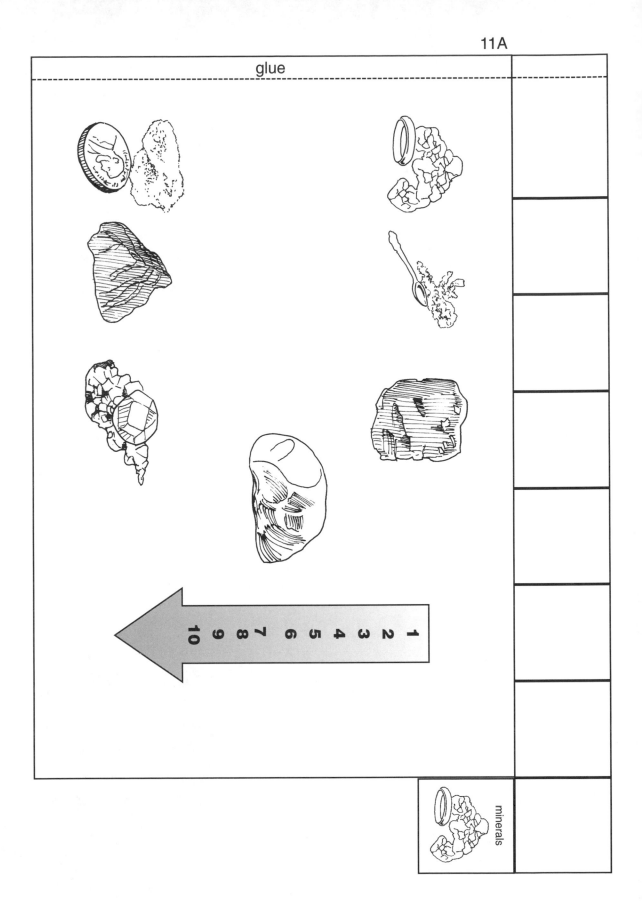

1 2 3 4 5 6 7 8 9 10

minerals

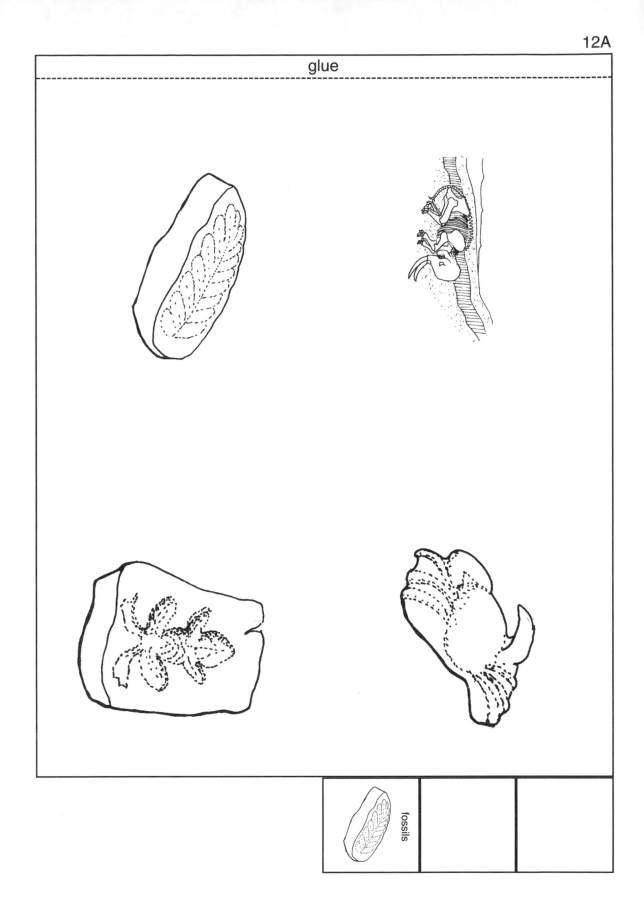

glue

fossils

glue

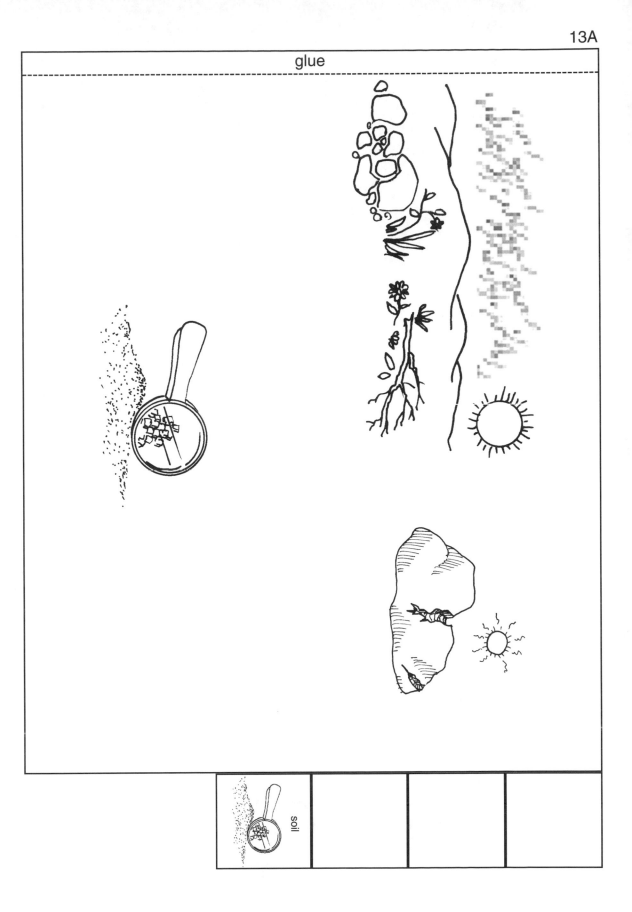

soil

glue

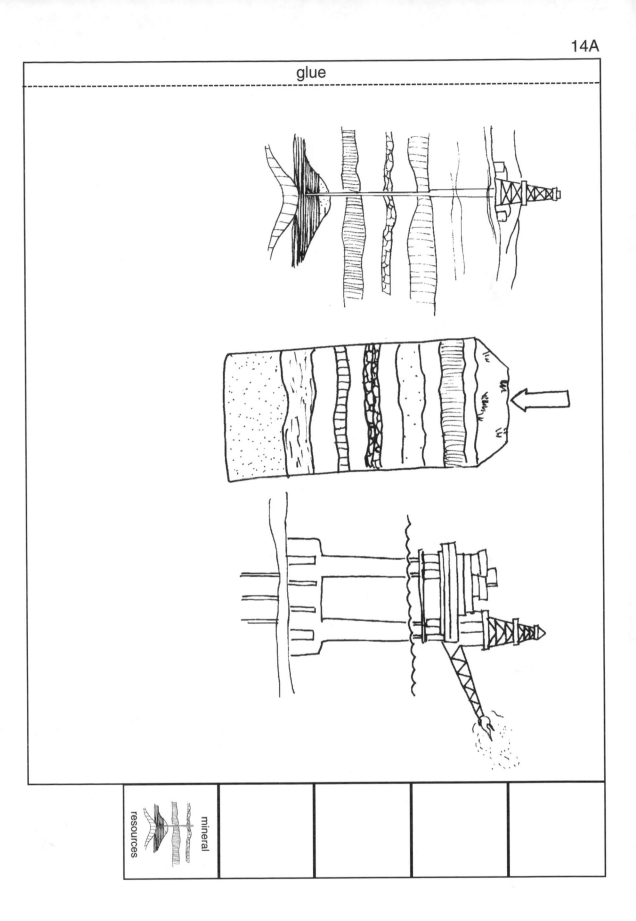

resources | mineral

glue

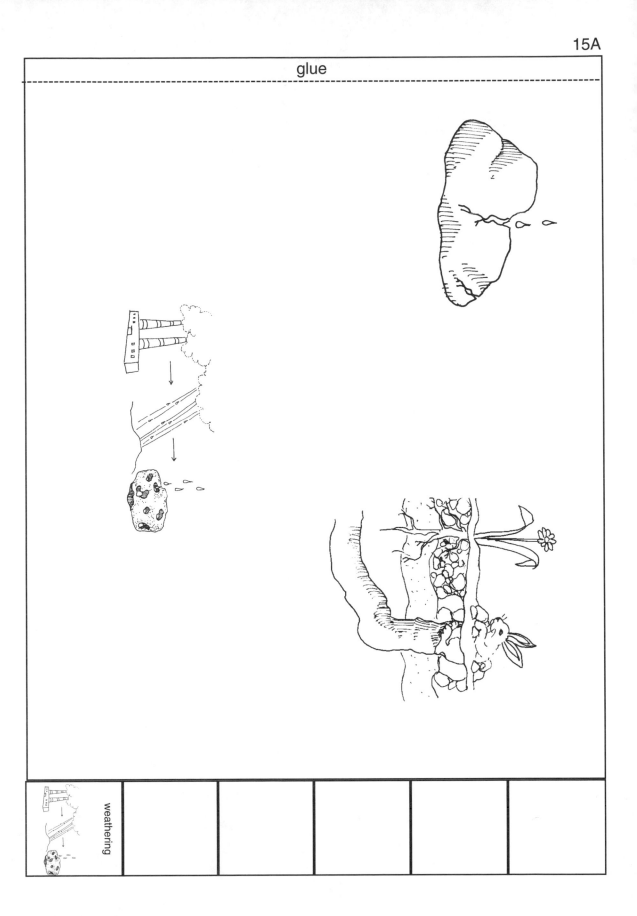

weathering

glue

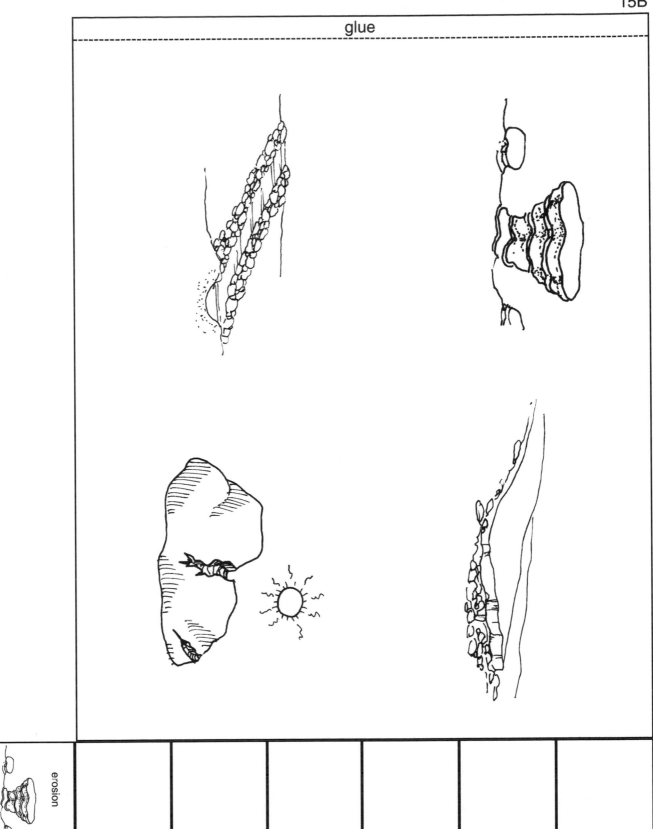

erosion

glue

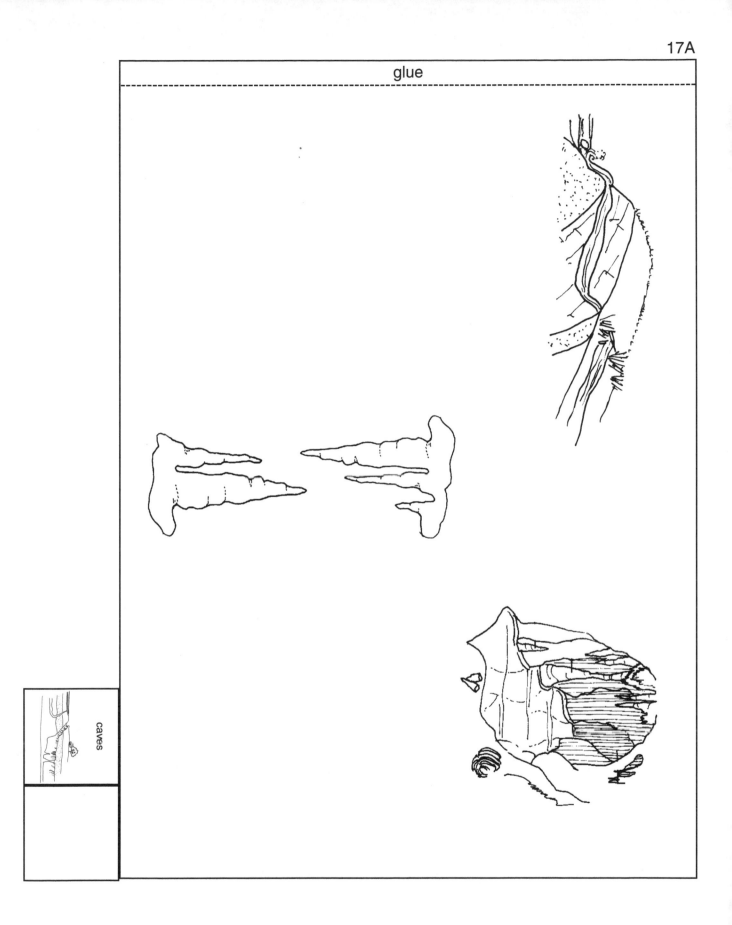

caves

glue

glaciers

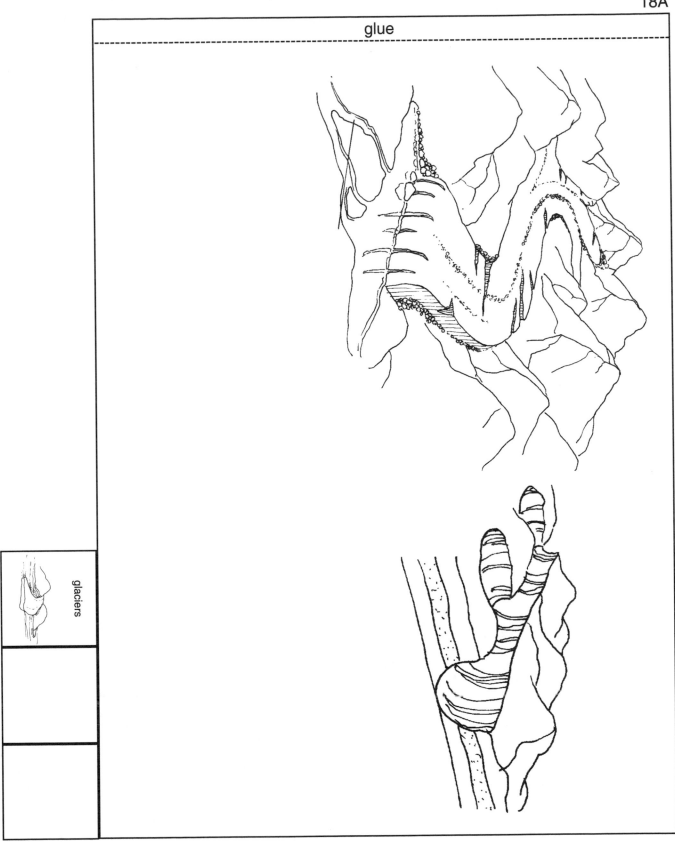

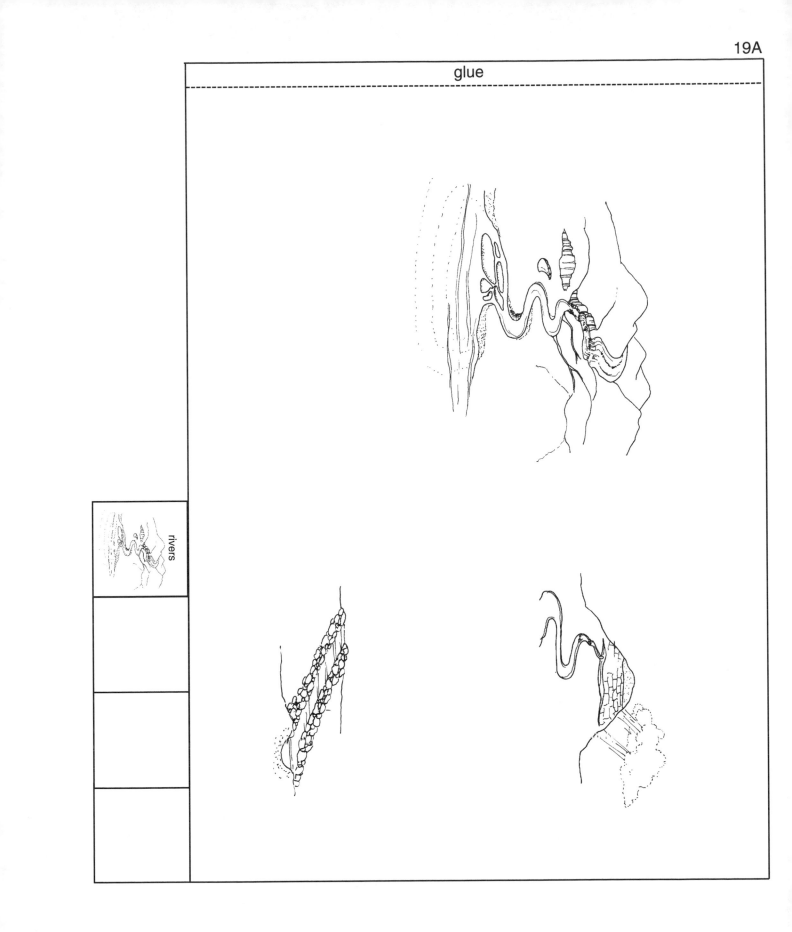

rivers

glue

river

features

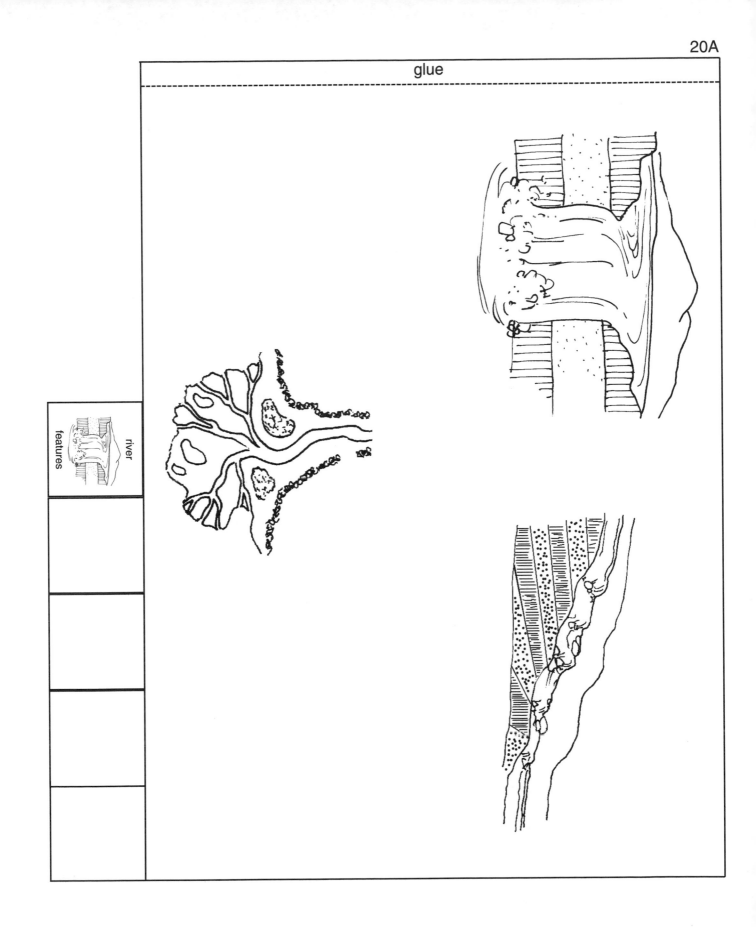

glue

lakes

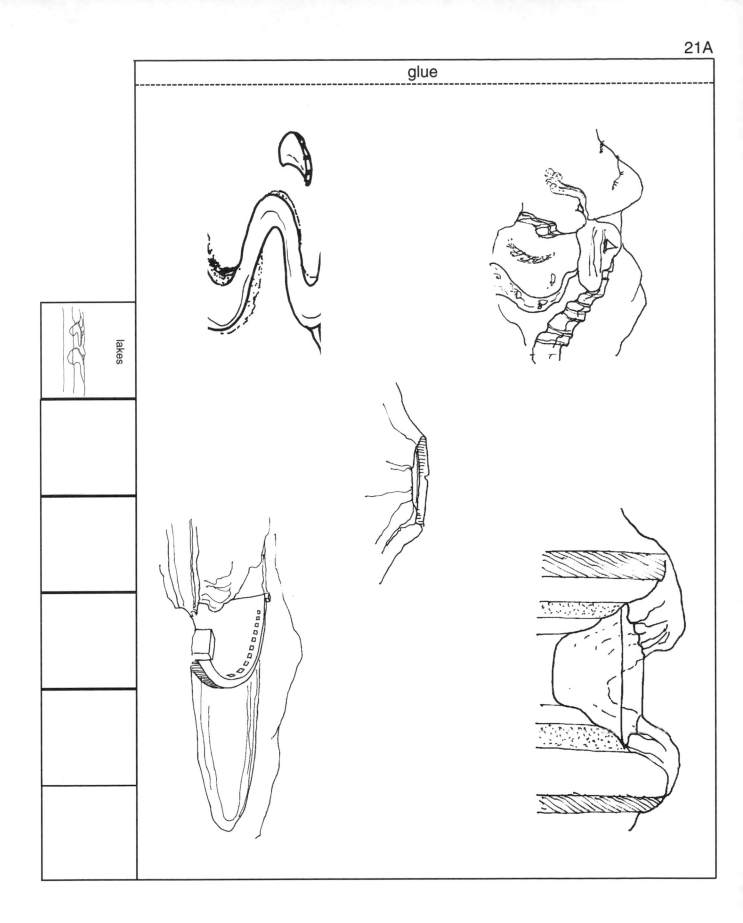

glue

coastlines

ocean floor

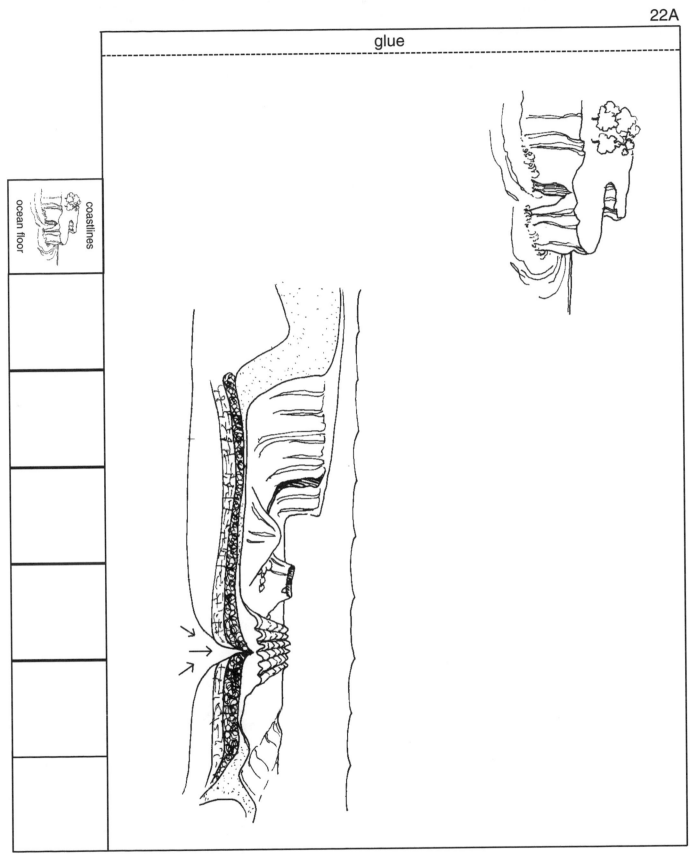

glue

deserts

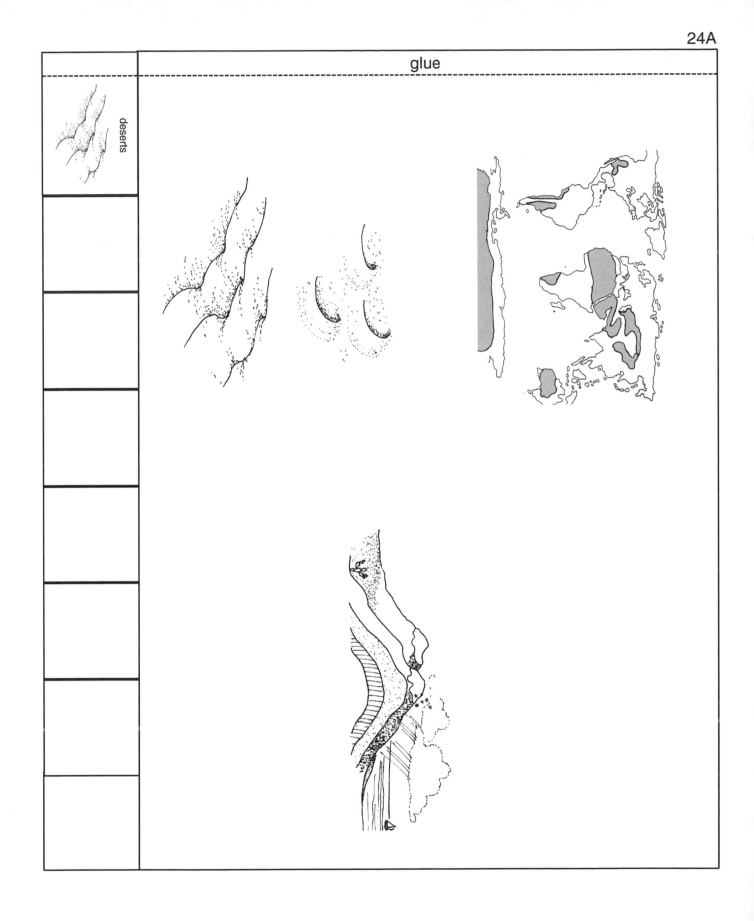